Discover Health

AGS

American Guidance Service, Inc.
Circle Pines, Minnesota 55014-1796
800-328-2560

Photos:

pp. 4, 90—Don Smetzer/Tony Stone Images; pp. 6, 186-87, 196, 199, 206—Michael Crousek; p. 10—Drew Tempe; pp. 16, 126—Lori Adamski Peek/Tony Stone Images; pp. 20-21, 106—Laura Dwight/Laura Dwight Photography; p. 22—Bill Bachmann/PhotoEdit; pp. 23, 54, 66—Jim Cummins/Tony Stone Images; p. 24—Bo Veisland, MI&I/Science Photo Library/Photo Researchers, Inc.; pp. 58, 164, 345, 368, 376, 384—Mary Kate Denny/PhotoEdit; pp. 60, 72, 120, 123, 136-37, 215, 265, 328—David Young-Wolff/PhotoEdit; pp. 68, 82-83—SuperStock International; p. 84—Mark Richards/PhotoEdit; pp. 85, 100, 118—Skjold Photography; pp. 86, 140, 333—David Kelly Crow/PhotoEdit; p. 93—D. MacDonald/PhotoEdit; pp. 96, 113, 160 bottom, 201, 218, 282, 289, 305, 312, 388—Tony Freeman/PhotoEdit; pp. 102, 262—Zigy Kaluzny/Tony Stone Images; p. 109—Penny Tweedie/Tony Stone Images; p. 138—Spencer Grant/PhotoEdit; pp. 139, 158, 173, 185, 212, 264, 296, 307, 347, 359, 378—Michael Newman/PhotoEdit; pp. 142, 147, 149—Felicia Martinez/PhotoEdit; pp. 143, 184, 280-81—Richard Hutchings/PhotoEdit; p. 145—Steven Needham/Envision; p. 150—Shaun Egan/TSI; p. 151—U.S. Dept. of Agriculture; p. 160 top—Chris Everard/Tony Stone Images; p. 182-83—David Hanover/Tony Stone Images; p. 192—Courtesy of Pathways Magazine and Toronto Public Health; pp. 228-29—Hubbard/Gamma Liaison International; p. 230—Robin L. Sachs/PhotoEdit; pp. 231, 242, 254—I. Burgum/P. Boorman/Tony Stone Images; p. 237—Robert Brenner/PhotoEdit; p. 246—AP/Wide World Photos; pp. 249, 255—Myrleen Ferguson/PhotoEdit; p. 256—John Bavosi/Science Photo Library/Photo Researchers, Inc.; p. 268—Hans-Ulrich Osterwalder/Science Photo Library/Photo Researchers, Inc.; p. 271—James Shaffer/PhotoEdit; pp. 283, 304, 322—Gary Holscher/Tony Stone Images; p. 285—Donald Johnston/Tony Stone Images; p. 297—David Weintraub/Photo Researchers, Inc.; p. 298—A. Ramey/PhotoEdit; p. 310—Courtesy of the American Red Cross. All Rights Reserved in all Countries, American Red Cross; p. 324—Jonathan Nourak/PhotoEdit; p. 342-43—David Joel/Tony Stone Images; p. 344—Robert E. Daemmrich/Tony Stone Images; p. 352—Chuck Keeler/Tony Stone Images; p. 355—Bruce Ayres/Tony Stone Images; p. 370—Jeff Greenberg/Unicorn Stock Photos; p. 387—Anne Heller; p. 390—Nello Giambi/Tony Stone Images; p. 394—David Woodfall/Tony Stone Images; p. 396—David Young-Wolff/Tony Stone Images; p. 398—Robert Cameron/Tony Stone Images; p. 409—1998 PhotoDisc; p. 413—Custom Medical Stock

©2000 AGS® American Guidance Service, Inc., Circle Pines, MN 55014-1796.
All rights reserved, including translation. No part of this publication may be reproduced or transmitted in any form or by any means without written permission from the publisher.

Printed in the United States of America

ISBN 0-7854-1843-1

Product Number 91050

A 0 9 8 7 6 5 4 3 2 1

Contents

How to Use This Book 10
Introduction 16

Unit 1 Personal Health and Family Life 20

Chapter 1 The Body Systems 23
 Lesson 1 Cells, Tissues, and Organs. 24
 Lesson 2 The Body's Protective Covering 26
 Lesson 3 Skeletal and Muscular Systems. 28
 Lesson 4 The Digestive and
   Excretory Systems 32
 Lesson 5 The Respiratory and
   Circulatory Systems. 36
 Lesson 6 The Nervous System 40
 Lesson 7 The Endocrine System. 44
 Lesson 8 The Reproductive System 47
   Chapter Summary and Review. 51

Chapter 2 Hygiene and Fitness. 54
 Lesson 1 Hygiene. 55
 Lesson 2 Fitness 59
   Chapter Summary and Review. 63

Chapter 3 The Family 66
 Lesson 1 The Family Life Cycle 67
 Lesson 2 Dealing With Family Problems 71
   Chapter Summary and Review. 75

Unit 2 **Mental and Emotional Health** 82

    Chapter 4   Emotions. 85
                    Lesson 1   Emotions and Their Causes 86
                    Lesson 2   Social Emotions 90
                    Lesson 3   Emotions and Behavior............. 93
                                Chapter Summary and Review....... 97

    Chapter 5   Maintaining Mental Health 100
                    Lesson 1   Managing Frustration 101
                    Lesson 2   Managing Stress and Anxiety 104
                    Lesson 3   Responding to Peer Pressure 108
                    Lesson 4   Eating Disorders 112
                                Chapter Summary and Review...... 115

    Chapter 6   Relationships 118
                    Lesson 1   Being a Friend to Yourself.......... 119
                    Lesson 2   Making Friends 122
                    Lesson 3   Healthy Relationships 125
                                Chapter Summary and Review...... 129

Unit 3 **Nutrition** **136**

Chapter 7   Diet and Health 139
　　　　　　Lesson 1   Food for Energy 140
　　　　　　Lesson 2   Carbohydrates and Protein 142
　　　　　　Lesson 3   Fats and Cholesterol 145
　　　　　　Lesson 4   Vitamins, Minerals, and Water 147
　　　　　　Lesson 5   Dietary Guidelines 151
　　　　　　　　　　　 Chapter Summary and Review 155

Chapter 8   Making Healthy Food Choices 158
　　　　　　Lesson 1   Food Choices and Health 159
　　　　　　Lesson 2   Influences on Food Choices 163
　　　　　　Lesson 3   Reading Food Labels 167
　　　　　　Lesson 4   Making Foods Safe 171
　　　　　　　　　　　 Chapter Summary and Review 175

Discover Health   *Contents*   **5**

Unit 4   Use and Misuse of Substances 182

Chapter 9   Medicines and Drugs. 185
  Lesson 1   Prescription and Over-the-Counter Medicines 186
  Lesson 2   The Effect of Medicines and Drugs on the Body 190
  Lesson 3   Tobacco 194
  Lesson 4   Alcohol 198
  Lesson 5   Narcotics, Depressants, Stimulants, and Hallucinogens 202
  Lesson 6   Other Dangerous Drugs 206
              Chapter Summary and Review 209

Chapter 10 Drug Dependence—Problems and Solutions. . . 212
  Lesson 1   The Problem of Drug Dependence . . 213
  Lesson 2   Solutions to Drug Dependence 217
              Chapter Summary and Review. 221

Unit 5   **Preventing and Controlling Diseases and Disorders ... 228**

    Chapter 11   Disease—Causes and Prevention............ 231
                Lesson 1   Causes of Disease................. 232
                Lesson 2   The Body's Protection
                            From Disease 235
                            Chapter Summary and Review 239

    Chapter 12   Preventing AIDS and Sexually Transmitted
                      Diseases................................ 242
                Lesson 1   AIDS 243
                Lesson 2   Sexually Transmitted Diseases 247
                            Chapter Summary and Review...... 251

    Chapter 13   Common Diseases......................... 254
                Lesson 1   Cardiovascular Disease 255
                Lesson 2   Cancer.......................... 259
                Lesson 3   Asthma 263
                Lesson 4   Diabetes 265
                Lesson 5   Arthritis 268
                Lesson 6   Epilepsy......................... 270
                            Chapter Summary and Review...... 273

Discover Health   *Contents*   **7**

Unit 6 **Injury Prevention and Safety Promotion** **280**

Chapter 14 Preventing Injuries 283
    Lesson 1 Promoting Safety 284
    Lesson 2 Reducing Risks of Fire 288
    Lesson 3 Safety for Teens 291
    Lesson 4 Emergency Equipment 295
    Lesson 5 Safety During Natural Disasters 297
    Chapter Summary and Review 301

Chapter 15 First Aid for Injuries 304
    Lesson 1 What to Do First 305
    Lesson 2 Caring for Common Injuries 307
    Lesson 3 First Aid for Bleeding, Shock, and Choking 311
    Lesson 4 First Aid for Heart Attacks and Poisoning 315
    Chapter Summary and Review 319

Chapter 16 Preventing Violence 322
    Lesson 1 Defining Violence 323
    Lesson 2 Causes of Violence 327
    Lesson 3 Preventing Violence 331
    Chapter Summary and Review 335

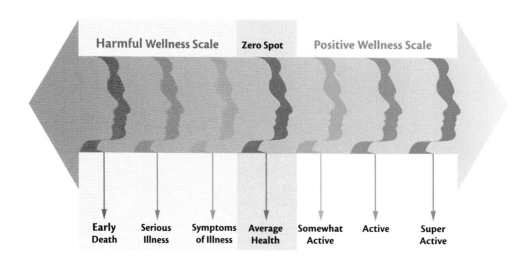

Unit 7   **Health and Society**................................**342**

    Chapter 17   Consumer Health 345
                    Lesson 1   Health Care Information 346
                    Lesson 2   Seeking Health Care 348
                    Lesson 3   Paying for Health Care 352
                    Lesson 4   Being a Wise Consumer 354
                    Lesson 5   Evaluating Advertisements 357
                    Lesson 6   Consumer Protection 361
                                    Chapter Summary and Review......365

    Chapter 18   Public Health 368
                    Lesson 1   Defining Community 369
                    Lesson 2   Community Health Resources......373
                    Lesson 3   Community Health Advocacy
                                      Skills 377
                                    Chapter Summary and Review......381

    Chapter 19   Environmental Health.....................384
                    Lesson 1   Health and the Environment 385
                    Lesson 2   Air Pollution and Health...........389
                    Lesson 3   Water and Land Pollution
                                  and Health 393
                    Lesson 4   Promoting a Healthy Environment..397
                                    Chapter Summary and Review......401

Appendices **408**

    Appendix A   Nutrition Charts and Tables 408
    Appendix B   Fact Bank................................410
    Glossary..414
    Index 422

How to Use This Book: A Study Guide

Welcome to the study of health. Everyone wants to have good health and wellness. Studying health helps you to learn ways to promote wellness. It helps you identify causes of health problems and ways to prevent them.

As you read the units, chapters, and lessons of this book, you will learn about promoting emotional, physical, and social health.

How to Study

- Plan a regular time to study.
- Choose a quiet desk or table where you will not be distracted. Find a spot that has good lighting.
- Gather all the books, pencils, and paper you need to complete your assignments.
- Decide on a goal. For example: "I will finish reading and taking notes on Chapter 1, Lesson 1, by 8:00."
- Take a five- to ten-minute break every hour to keep alert.
- If you start to feel sleepy, take a short break and get some fresh air.

Before Beginning Each Unit

- Read the title and the opening paragraph.
- Study the photograph. What does the photo say to you about health?
- What does the quotation say to you?
- Read the titles of the chapters in the unit.
- Read "What Do You Think?" and consider what you would do in that situation.
- Look at the headings of the lessons and paragraphs to help you locate main ideas.
- Read the chapter and unit summaries to help you identify key issues.
- Read the Deciding for Yourself page at the end of each unit.

Each unit covers a different health topic.

How to Use This Book: A Study Guide

Before Beginning Each Chapter

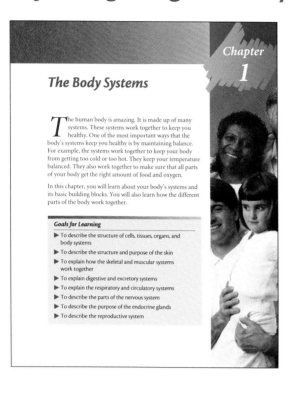

- Read the chapter title.
- Study the goals for learning. The chapter review and tests will ask questions related to these goals.

Before Beginning Each Lesson

Read the lesson title and restate it in the form of a question. For example:

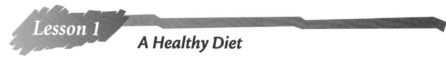

A Healthy Diet

Write: *What is a healthy diet?*

Look over the entire lesson, noting . . .

- pictures
- tables
- charts
- figures
- bold words
- text organization
- questions in the margins
- lesson review

Also note these features . . .

- Action for Health—An action you can take
- Careers—A health career and its requirements
- Healthy Subjects—A subject such as math or literature related to the chapter topic
- Technology—A technology advance related to the chapter topic
- Then and Now—An explanation of how health was approached in earlier years compared with today

- Tips—A short, easy-to-use tip on health, fitness, safety, or nutrition
- Writing About Health—Write about how a topic applies to your health and life

As You Read the Lesson

- Read the major headings. Each subhead is a question.
- Read the paragraphs that follow to answer the question.
- Before moving on to the next heading, see if you can answer the question. If you cannot, reread the section to look for the answers. If you are still unsure, ask for help.
- Answering the questions in the lesson will help you determine if you know the lesson's key ideas.

Using the Bold Words

Knowing the meaning of all the boxed words in the left column will help you understand what you read.

These words appear in **bold type** the first time they appear in the text and are defined in the paragraph.

> Proteins are made up of smaller units called **amino acids.**

All of the words in the left column are also defined in the **glossary.**

> **Amino acids**—The smaller units of protein (p. 144)

Bold type
Words seen for the first time will appear in bold type

Glossary
Words listed in this column are also found in the glossary

Taking Notes in Class

Some students prefer taking notes on index cards.

Others jot down key ideas in a spiral notebook.

As you read, you will be learning many new facts and ideas. Your notes will be useful and will help you remember when preparing for class discussions and studying for tests.

- Always write the main ideas and supporting details.
- Use an outline format to help save time.
- Keep your notes brief. You may want to set up some abbreviations to speed up your note-taking. For example: *with = w/ and = + dollars = $*
- Use the same method all the time. Then when you study for a test, you will know where to find the information you need to review.

Here are some tips for taking notes during class discussion:

- Use your own words.
- Do not try to write everything the teacher says.
- Write down important information only.
- Don't be concerned about writing in complete sentences. Use phrases.
- Be brief.
- Rewrite your notes to fill in possible gaps as soon as you can after class.

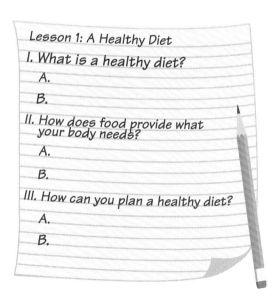

Using an Outline

You may want to outline the section using the subheads as your main points. An outline will help you remember the major points of the section. An example of an outline is shown at left. Your teacher may have you use the Student Study Guide for this book.

Getting Ready to Take a Test

The Summaries and Reviews can help you get ready to take tests.

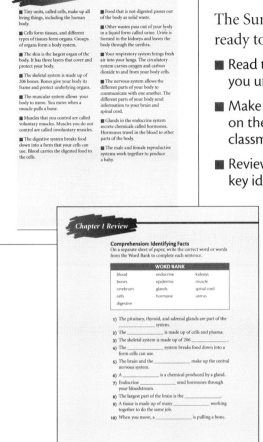

- Read the summaries from your text to make sure you understand the chapter's main ideas.
- Make up a sample test of items you think may be on the test. You may want to do this with a classmate and share your questions.
- Review your notes and test yourself on words and key ideas.
- Practice writing about some of the main ideas from the chapter.
- Answer the questions under Identifying Facts.
- Answer the questions under Understanding the Main Ideas.
- Write what you think about the questions under Write Your Opinion.

Use the Test Taking Tip

- Read the Test Taking Tip with each Chapter Review of the text.

Test Taking Tip If you know you will have to define certain terms on a test, write the term on one side of a card. Write its definition on the other side. Use the cards to test yourself, or work with a partner.

Introduction

Good health is basic to everything in life. Without health, people have difficulty doing the things they want to do. But health is more than being without illness. Health is a state of being healthy. There are three parts of good health.

- *Physical health* is the body's ability to meet the demands of daily living. It allows a person to do things without getting tired and run-down. For example, it allows you to attend school, to study, and to participate in after-school activities.

- *Social health* is the ability to get along with other people. It is the ability to make friends and to be a friend. Social health is the ability to work with others and to perform effectively as part of a group. It involves contributing to the good of the family and helping out in the community.

It involves trust, honesty, respect, responsibility, fairness, and good citizenship.

- *Emotional health* is the ability to handle the problems and pressures of daily living. It also involves feeling good about oneself. It involves controlling and managing anger without turning to violence.

The physical, social, and emotional parts of health are all connected. If people are physically ill, it affects how they feel about themselves. Physical illness affects the ability to relate well with other people. In a similar way, emotional problems can contribute to physical illnesses such as heart disease. For example, you may feel extra tired when you have pressures at home or at school. You may notice you are more likely to have accidents or to get upset when you are worried about something.

How Is Good Health Achieved?

Good health is something that everyone must achieve for himself or herself. We all should work toward promoting good physical, social, and emotional health and preventing illness. Like playing a musical instrument well, good health requires practicing certain basic rules for keeping healthy. For example, everyone needs to practice developing good friendships. We all need to practice handling our feelings appropriately. Good health requires that we use good judgment and self-control to provide the body with the food, exercise, rest, and care it needs.

- **Exercise regularly.**
- **Eat three balanced, healthy meals daily, including breakfast.**
- **Choose to eat only healthy snacks.**
- **Maintain a normal weight for your height and age.**
- **Sleep eight hours each night.**
- **Do not smoke, drink alcohol, or use drugs that a doctor hasn't directed you to take.**
- **Get reliable medical and dental care and advice.**

Achieving and maintaining good health is largely within our control. By practicing the healthy behaviors listed in the box, we can help increase the quality and length of our life.

All of these behaviors are a way of preventing health problems and achieving wellness.

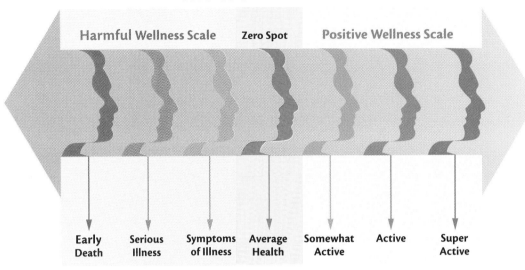

What Is Wellness?

Wellness is an active state of health in which an individual moves toward balancing physical, social, and emotional health. The Wellness Scale shows how to chart a person's level of wellness. At one end of the scale is a high state of physical, social, and emotional health. People in the middle of the scale have average health. At the other end of the scale is serious illness and early death.

As you read this textbook, you will learn more about what is involved in achieving good health and wellness. But before you begin, think about your present health and its three parts. Think about where you are on the Wellness Scale. To help you rate your health and wellness, use the Health Self-Rating Chart to rate some behaviors and choices that may affect your health. The chart may help you find areas of your health that you would like to improve.

HEALTH SELF-RATING CHART

Directions: Number a sheet of paper to correspond to each chart section. Read each statement. If the statement is mostly true for you, write *yes*. If it is mostly false, write *no*. You do not need to share your results with anyone. Score each section as the instructions tell you. To score your overall health level, add your scores on all sections.

26–30 = top 10 percent
21–25 = above average
16–20 = average
15 and lower = below average

PHYSICAL HEALTH

1. I eat a healthy breakfast every day.
2. I eat a balanced diet that is different every day.
3. I avoid unhealthy snacks.
4. I do not use tobacco or alcohol or take drugs that a doctor hasn't directed me to use.
5. I get eight hours of sleep each night.
6. I take part in exercises and sports I like.
7. I work at developing muscle tone and fitness three times a week.
8. I bike, swim, run, or walk for at least thirty minutes three or more times a week.
9. I relax at least ten minutes a day.
10. I get regular medical and dental checkups.

Score one point for each *yes* answer.
  9–10 = top 10 percent
  7–8 = above average
  5–6 = average
  4 or lower = below average

SOCIAL HEALTH

1. I meet people and make friends often.
2. I have one or two close friends.
3. I can say "no" to my friends if they want me to do something I don't want to do.
4. I balance having my way with letting others have their way.
5. I respect other people's right to be different from me.
6. I am able to work cooperatively with others.
7. If I have a problem with other people, I face the problem and try to work it out with them.
8. I am comfortable communicating with adults.
9. I am comfortable talking with both females and males my age.
10. I am fair and trustworthy with others.

Score one point for each *yes* answer.
  9–10 = top 10 percent
  7–8 = above average
  5–6 = average
  4 or lower = below average

EMOTIONAL HEALTH

1. I try to accept my feelings of love, fear, anger, and sadness.
2. I can tell when I am under pressure.
3. I try to find ways to deal with pressure and control it.
4. I try to have a positive outlook.
5. I ask for help when I need it.
6. I have friends and relatives with whom I discuss problems.
7. I give compliments.
8. I can accept and use helpful comments.
9. I take responsibility for my actions.
10. I am honest with myself and others.

Score one point for each *yes* answer.
  9–10 = top 10 percent
  7–8 = above average
  5–6 = average
  4 or lower = below average

Introduction

"Health is a state of complete physical, mental, and social well-being, and not merely the absence of disease..."
—World Health Organization Constitution

Unit 1

Personal Health and Family Life

List the different things you do in one day. Your list might have walking, reading, laughing, and playing sports on it. Whatever you do, think, or feel—it all happens because your body and all its parts work together. Your body works all the time. It works even when you are asleep.

Good health doesn't just happen. In this unit, you will see what kinds of things you can do to keep yourself healthy. You will learn about the different parts of your body and how they work together. You will learn what you can do to keep your body and its parts working well. You will also learn that your family is important to your health.

▶ Chapter 1 The Body Systems
▶ Chapter 2 Hygiene and Fitness
▶ Chapter 3 The Family

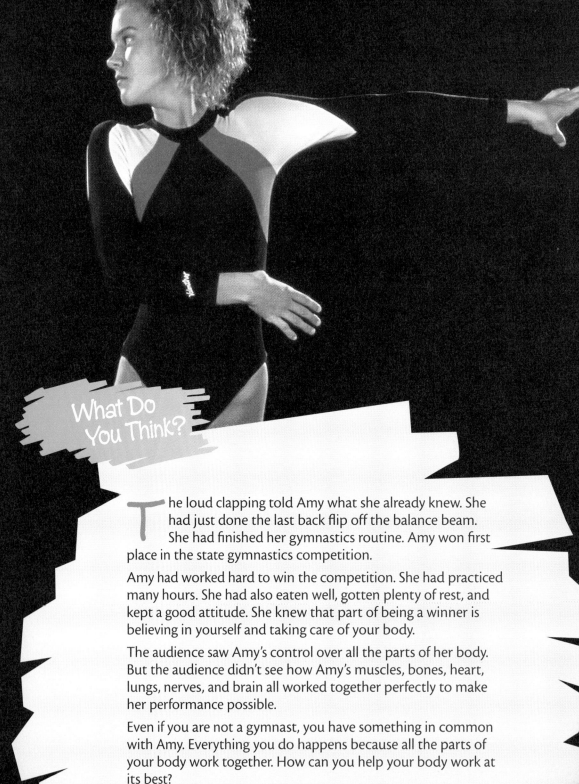

What Do You Think?

The loud clapping told Amy what she already knew. She had just done the last back flip off the balance beam. She had finished her gymnastics routine. Amy won first place in the state gymnastics competition.

Amy had worked hard to win the competition. She had practiced many hours. She had also eaten well, gotten plenty of rest, and kept a good attitude. She knew that part of being a winner is believing in yourself and taking care of your body.

The audience saw Amy's control over all the parts of her body. But the audience didn't see how Amy's muscles, bones, heart, lungs, nerves, and brain all worked together perfectly to make her performance possible.

Even if you are not a gymnast, you have something in common with Amy. Everything you do happens because all the parts of your body work together. How can you help your body work at its best?

Chapter 1

The Body Systems

The human body is amazing. It is made up of many systems. These systems work together to keep you healthy. One of the most important ways that the body's systems keep you healthy is by maintaining balance. For example, the systems work together to keep your body from getting too cold or too hot. They keep your temperature balanced. They also work together to make sure that all parts of your body get the right amount of food and oxygen.

In this chapter, you will learn about your body's systems and its basic building blocks. You will also learn how the different parts of the body work together.

Goals for Learning

▶ To describe the structure of cells, tissues, organs, and body systems
▶ To describe the structure and purpose of the skin
▶ To explain how the skeletal and muscular systems work together
▶ To explain digestive and excretory systems
▶ To explain the respiratory and circulatory systems
▶ To describe the parts of the nervous system
▶ To describe the purpose of the endocrine glands
▶ To describe the reproductive system

Lesson 1

Cells, Tissues, and Organs

Cell
The basic unit that makes up your body

Cell membrane
The outer wall of a cell

Cytoplasm
The jelly-like material inside the cell membrane

Microscope
A tool used to see cells

Nucleus
The control center of the cell

Look at your skin. What do you see? You may see lines on the skin or flakes of skin. What you cannot see just by looking at your skin are the tiny **cells** that make up your skin. In fact, tiny living units, called cells, make up each part of your body.

What Are Cells?

Cells are the basic units that make up all living things. Your body is made up of trillions of cells. Cells are too small to see without a tool called a **microscope**. A microscope can make a cell look bigger than it really is, so you can see what it looks like. If you were to look at a cell using a microscope, it might look like the cell shown in Figure 1.1.

Your body is made up of many different kinds of cells. Each kind of cell has its own job. For example, skin cells help keep germs out of your body. White blood cells fight the germs that do enter your body. Nerve cells send messages among the different parts of your body and your brain. Each kind of cell carries out a different job.

What Are the Main Parts of a Cell?

Look again at the picture of the cell in Figure 1.1. Most cells have three basic parts: a **cell membrane**, the **cytoplasm**, and a **nucleus**.

The cell membrane is the outer wall of the cell. It is like a thin skin around the cell. Food and oxygen can pass into the cell and wastes can pass out of the cell through the cell membrane. The cytoplasm is a jelly-like liquid inside the cell membrane. The cytoplasm often has other cell parts floating in it. Most of the cell's life activities take place in the cytoplasm. The nucleus is sometimes called the control center of the cell. It has all the information that the cell needs to carry out its job and to make new cells like itself.

Figure 1.1. A typical cell

Body system
A group of organs that work together to carry out a certain job

Mitosis
The dividing process that makes new cells

Organ
A group of tissues that work together

Tissue
A group of cells that do the same job

The nucleus makes new cells by splitting into two, making one cell two cells. When the two cells split, they become four cells, which split to make eight. This dividing is called **mitosis**. Mitosis allows your body to make new cells to grow or to replace dead cells.

What Is a Body System?

Cells that do the same job combine to form a **tissue**. Muscles and bones are examples of tissues. Different types of tissues working together form an **organ**, such as the stomach or the kidneys. A group of organs working together to carry out a certain job is called a **body system**. The digestive system is an example of a body system. Each system carries out its own job and works with other systems to keep the body healthy.

LESSON 1 REVIEW Write the answers to these questions on a separate sheet of paper. Use complete sentences.

Why are cells called "building blocks of life"?

1) Name three different kinds of cells and what they do.
2) Why is a microscope needed to look at cells?
3) How does mitosis help you grow?
4) How is tissue formed?
5) What are the three main parts of a cell, and what does each one do?

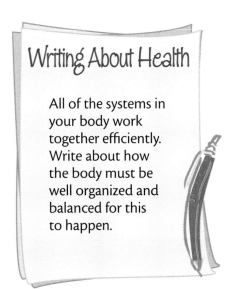

Writing About Health

All of the systems in your body work together efficiently. Write about how the body must be well organized and balanced for this to happen.

Lesson 2

The Body's Protective Covering

Dermis
The middle layer of the skin

Epidermis
The outer layer of skin that you can see

Gland
A group of cells that produces a special substance to help the body work

Keratin
A protein that makes nails hard

Melanin
A chemical that gives skin and hair its color

Perspiration
Sweat

Your skin is the largest organ of your body. It covers and protects your body.

What Does Your Skin Do?

The skin protects your body by preventing germs and other harmful elements from getting inside. Your skin also keeps your body temperature steady. Your skin has more than four million pores, or tiny holes, which help cool your body. A **gland** is a group of cells that produces a special substance to help the body work. For example, if you get too hot, sweat glands in your skin give off moisture called sweat, or **perspiration**. Perspiration comes to the surface of your body through the pores in the skin. As perspiration evaporates from the surface of your skin, your body cools.

Your skin also helps your body adjust to changes around it. Skin detects pressure, pain, heat, and cold. This information is then sent to your brain. The brain interprets the messages and tells different parts of your body how to react.

What Is the Structure of the Skin?

Your skin has three main layers. You can see these layers in Figure 1.2. The outer layer, called the **epidermis**, is the part of the skin that you see. It is made up of a thin layer of dead skin cells. Your body is always making new skin cells. It takes about twenty-eight days for new skin cells to form and move to the outer layer of your skin. As the cells travel to the outer layer of skin, they harden and die. When they get to the outside of your body, the dead cells rub off, leaving behind the younger cells. The color of your skin and hair comes from a substance in the epidermis called **melanin**.

Some of the cells in the epidermis develop large amounts of a protein called **keratin**. Keratin hardens into nails, which protect the ends of your fingers and toes.

The second layer of skin is the **dermis**. It is made of living cells. The dermis has blood vessels, nerves, and glands. New

Why do you think animals have so many different kinds of skin?

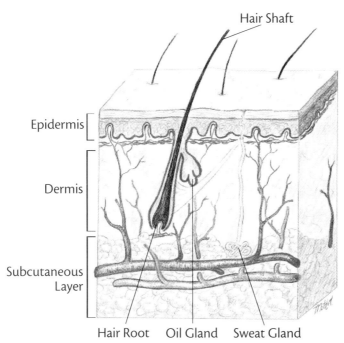

Figure 1.2.
Cross section of the skin

Subcutaneous layer
The deepest layer of the skin

skin cells are also made in the dermis. Oil glands in the dermis help keep your skin smooth and able to stretch. Sweat glands secrete perspiration. Your hair grows from hair follicles in the dermis.

The **subcutaneous layer** of the skin is closest to your bones. It connects your skin to your muscles. This layer is made of fatty tissue. The fat helps protect your body from very hot or very cold temperatures. It also protects your body when you bump into things.

LESSON 2 REVIEW Write the answers to these questions on a separate sheet of paper. Use complete sentences.

1) What are two ways the skin protects your body?
2) What are the three main layers of the skin?
3) What are the similarities and differences between the epidermis and the dermis?
4) What are two types of glands in the skin? What do they do?
5) How does skin respond to changes, such as pain or heat?

Lesson 3

The Skeletal and Muscular Systems

Contract
Shorten

Joint
A place where two bones come together

Muscular system
The body system made up of your muscles

Skeletal system
The body system made up of your bones and joints

Two body systems that give your body its shape and help it to move are the **skeletal system** and the **muscular system**. These two systems work together and with other systems to keep you healthy.

What Is the Skeletal System?

Figure 1.3 shows the skeletal system. It is made up of 206 bones. The places where each bone connects to another bone are called **joints**. Tough bands of tissue hold joints together.

Your skeletal system provides the frame for the rest of your body. Your bones are hard. This hardness provides the frame for your body. Most of the other parts of your body are soft.

Your skeletal system also protects other parts of your body. For example, your ribs form a cage around your heart and lungs. This cage protects your heart and lungs from being crushed when you catch a football or hug someone. The bones of your head protect your brain when you bump your head.

Your skeletal system also allows your body to stand up and to walk and move around. But, in order to move, your skeletal system must work closely with your muscular system.

What Is the Muscular System?

Bones cannot move by themselves. They are moved by the muscles that make up your muscular system. Muscles move bones by becoming shorter, or **contracting**. When a muscle that is attached to a bone contracts, it pulls the bone. Muscles move bones in only one direction. This is why muscles work in pairs. For example, one set of muscles works to raise your arm up while another set is used to lower it.

Fitness Tip

Instead of watching TV, do something physical—walk, bike, swim, or run.

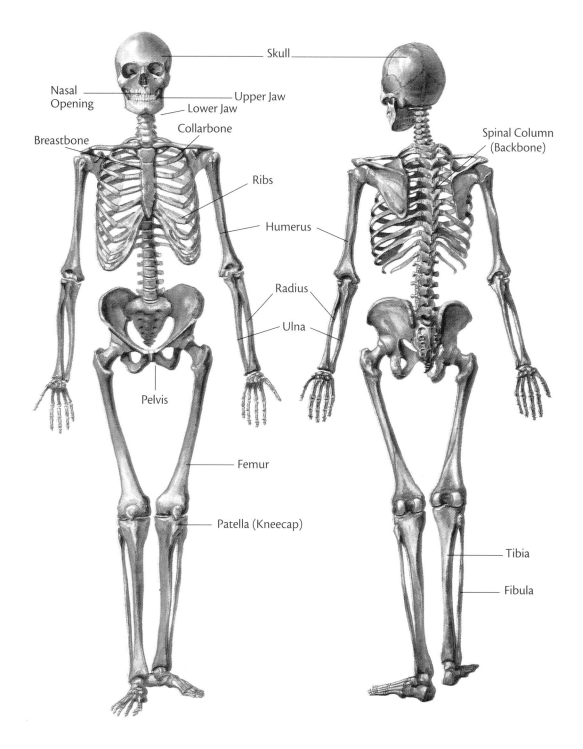

Figure 1.3. The skeletal system (front and back views)

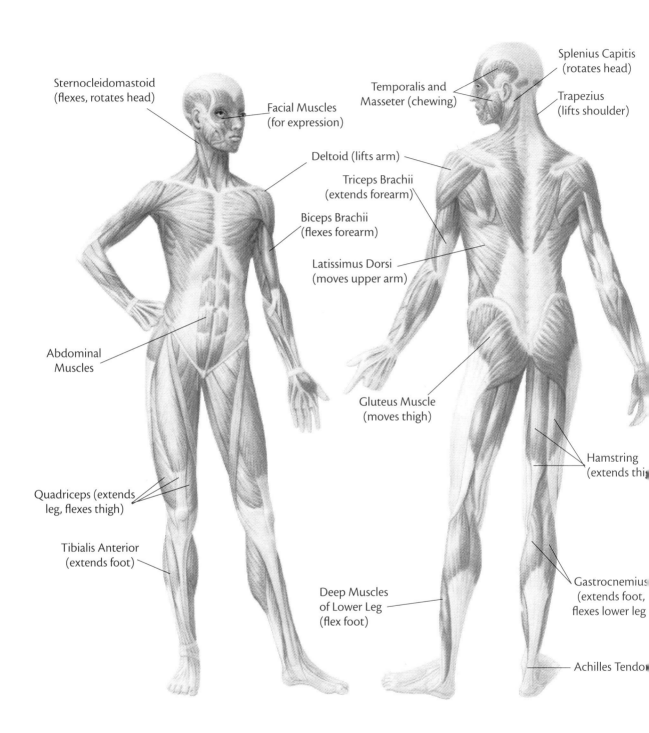

Figure 1.4. The muscular system (female front view and male back view)

30 Chapter 1 The Body Systems

Healthy Subjects: Consumer Science

RUNNING SHOES

Many people choose running as a way to keep fit. Choosing the right shoes is important for healthy feet. For most runners, "stability shoes" offer good cushioning, center support, and durability. However, "motion-control shoes" are better for people who tend to roll their feet inward as they step. These shoes are heavier but provide much-needed support for runners. Other runners' feet may be less flexible in the center. These people need cushioned shoes without much middle support. Off-road runners should wear "trail shoes." The soles of trail shoes give extra traction. They also include toe "bumpers," durable tops, and stronger stitching. The next time you buy shoes, consider how to keep your feet happy.

Involuntary muscle
A muscle that moves whether you think about it or not

Voluntary muscle
A muscle that moves when you think about it

What Kinds of Muscles Are There?

Your body has two basic kinds of muscles. They are called **voluntary muscles** and **involuntary muscles**. When you want to take a walk, your voluntary muscles cause you to walk. You control the movements that are caused by voluntary muscles.

Involuntary muscles cause movements whether you think about them or not. For example, your heart muscle is an involuntary muscle. It pumps blood all the time without you thinking about it. Voluntary muscles move your skeletal and muscular system. For example, voluntary muscles move your arms or legs as you walk and talk.

Why do you think it hurts to move after a muscle has been injured?

LESSON 3 REVIEW Write the answers to these questions on a separate sheet of paper. Use complete sentences.

1) How many bones make up the skeletal system?
2) How do the skeletal and muscular systems work together?
3) What is the purpose of the skeletal system?
4) Give an example of an involuntary and a voluntary muscle.
5) Explain how the muscle pairs in your arm allow you to lift a glass of milk from a table.

Lesson 4: The Digestive and Excretory Systems

Digestive system
The body system that breaks food down

Esophagus
The tube that connects the throat and the stomach

Excretory system
The body system that rids the body of waste and extra water

Saliva
The liquid in the mouth that begins digestion

Small intestine
Where most of digestion takes place

Your body needs food to live and grow. It also has to rid itself of wastes. The **digestive system** works to break food down into a form that your cells can use. The **excretory system** rids your body of its wastes.

What Does the Digestive System Do?

All the cells in your body use the food you eat to live, grow, and heal. First, the food must be broken down into a form that your cells can use. Breaking food down into a usable form is the job of your digestive system.

Look at the picture of your digestive system in Figure 1.5. Notice that your digestive system is made up of many organs that work together to break down food.

What Is the Path of Food Through the Body?

As you read, trace the path of food through the picture in Figure 1.5. Food enters your body through your mouth. In your mouth, teeth begin breaking the food down by chewing. **Saliva**, the liquid in your mouth, wets food and makes it easier to swallow. Chemicals in your saliva also begin breaking down the food. Your tongue pushes food to the back of your mouth. Then muscles force the food down your throat and through a tube, called the **esophagus**, into your stomach. Your stomach mixes the food with its own chemicals. Like the chemicals in your saliva, these chemicals also change the food into a form that your cells can use.

From the stomach, the food moves to the **small intestine**. In the small intestine, your blood takes in the digested parts of the food. The blood carries the food parts to every cell in your body.

Anus
The opening through which solid wastes leave the body

Large Intestine
The tube that connects the small intestine and the rectum

Rectum
The organ that stores solid waste before it leaves the body

How Does the Body Rid Itself of Solid Wastes?

The parts of the food that cells cannot use are wastes. These wastes pass to the **large intestine**. The large intestine is a tube-like organ connected to the small intestine. In the large intestine, most of the water that is mixed with the undigested food returns to the blood. The remaining waste is solid. Solid waste is stored in the **rectum**. Then it passes out of the body through the **anus**.

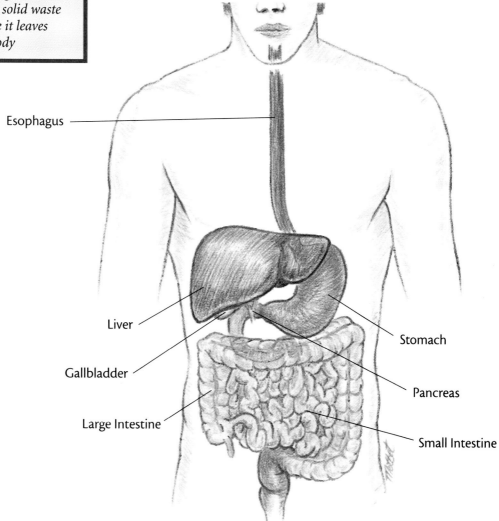

Figure 1.5. The digestive system

The Body Systems Chapter 1

Ureter
The tube that carries urine from the kidney to the bladder

Urethra
The tube through which urine passes out of the body

Urine
The liquid waste formed in the kidneys

How Does the Body Rid Itself of Liquid Wastes?

The excretory system also rids the body of waste products along with extra water. The wastes that leave the body through the excretory system are in the form of **urine**.

The excretory system is made up of two kidneys, two **ureters**, the bladder, and the **urethra**. These organs are shown in Figure 1.6.

Nutrition Tip

Drinking at least six to eight 8-ounce glasses of water daily will help the kidneys function properly.

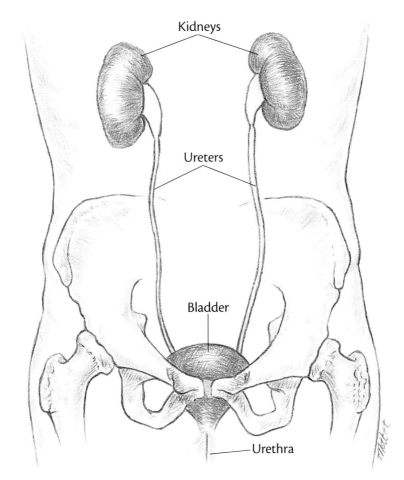

Figure 1.6. The excretory system

34 Chapter 1 The Body Systems

Then and Now

GINGER ISN'T JUST FOR COOKING

Many old herbal treatments are now considered useless or even harmful. However, some herbal treatments are now considered useful. For example, about 2,500 years ago, the Chinese used ginger to treat upset stomach, coughing, vomiting, and other ailments. Ginger was used by other groups, too. Nigerians have used it to treat yellow fever and malaria. In Russia, ginger has been used to treat toothaches.

Scientists have found that a chemical in ginger makes it a useful treatment for nausea, vomiting, migraines, and rheumatoid arthritis. Many pregnancy manuals and doctors who care for pregnant women advise taking ginger teas or ginger ale to help with morning sickness. Ginger is also a common treatment for motion sickness.

What conclusions can you make about what you eat and how it affects your digestive system?

At the same time that your blood provides food to your cells, it also picks up wastes from your cells. These wastes are processed in your kidneys. Your kidneys are on either side of your backbone. As blood flows through the kidneys, wastes from the blood are absorbed. Water and substances that your body can use are returned to your blood. Inside the kidneys, the wastes become urine. The urine moves from each kidney through one of the ureters to your bladder. Finally, urine leaves the body through the urethra, which leads outside the body.

LESSON 4 REVIEW Write the answers to these questions on a separate sheet of paper. Use complete sentences.

1) Where is food first broken down in the body?
2) What part of digestion takes place in the small intestine?
3) What happens to the parts of food that are not used by cells?
4) What waste product is produced by the excretory system?
5) Why is it important to have your kidneys work properly?

Lesson 5

The Respiratory and Circulatory Systems

Capillary
A tiny blood vessel

Circulatory system
The body system that pumps blood through the body

Respiratory system
The body system responsible for breathing

Trachea
The tube that connects the throat to the lungs

When you breathe, you take in oxygen and give off carbon dioxide. Your cells need oxygen to live. They also need a way to rid themselves of their carbon dioxide waste. The **circulatory system** and the **respiratory system** do this job together. They bring oxygen to your cells and take carbon dioxide away from your cells.

What Does the Respiratory System Do?

Your respiratory system brings fresh air that is rich in oxygen into your body. It also takes air filled with carbon dioxide out of your body.

Look at the picture of the respiratory system in Figure 1.7. When you breathe in air, it passes through your nose or your mouth into your body. The air you breathe in is cleaned and warmed as it passes through your nose.

From your nose or mouth, the air flows into your throat and down the windpipe, or **trachea**. Near your lungs, the trachea splits into two tubes. One tube leads to your left lung. The other tube leads to your right lung. These tubes bring the air to your lungs. Like a tree trunk, each tube divides into many branches that are smaller and smaller. The smallest branches lead into clusters of tiny air sacs.

How Does Oxygen Get to the Rest of the Body?

The air sacs lie next to tiny blood vessels, called **capillaries**. Both air sacs and capillaries have very thin walls. Oxygen moves through the thin walls of the air sacs into the capillaries.

Blood in the capillaries carries the oxygen from your lungs to all the cells in your body. At the same time, the blood picks up carbon dioxide from the cells. The carbon dioxide passes through the walls of the capillaries and into the air sacs. When you breathe out, your body gets rid of the carbon dioxide waste.

Why do you think your breathing increases when you exercise?

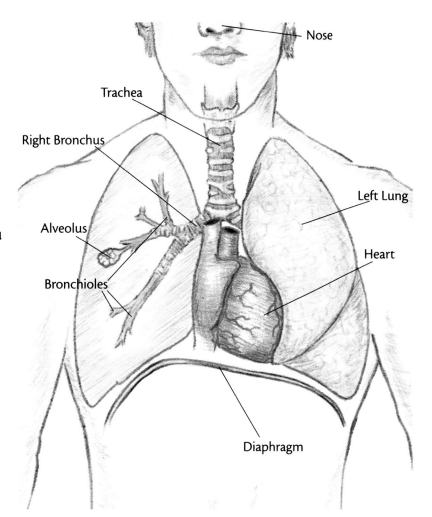

Figure 1.7. The respiratory system

What Is the Circulatory System?
The main job of the circulatory system is to pump blood through your entire body. The heart and blood vessels are the main organs that carry out this job.

What Does the Blood Do?
Blood does many jobs in the body. It brings food and oxygen to all parts of the body. It picks up wastes from all parts of the body. Blood also has cells that fight germs in the body. These cells help prevent illnesses.

Aorta
The largest artery in the body

Artery
A blood vessel that carries blood away from the heart

Plasma
The liquid part of blood

Vein
A blood vessel that carries blood back to the heart

Your blood is made of cells and a liquid, called **plasma**. Blood cells include white blood cells and red blood cells. White blood cells fight germs. Red blood cells carry oxygen from your lungs to the cells in your body. The red blood cells exchange the oxygen for carbon dioxide. Then they carry the carbon dioxide back to the lungs.

What Are the Parts of the Circulatory System?

Figure 1.8 shows the parts of the circulatory system. Your heart is not much bigger than your fist. However, it is a strong muscle that pumps blood to every part of your body. Your heart works nonstop, day and night, for your entire life.

Notice that the circulatory system is made up of many blood vessels. Blood vessels that carry blood away from your heart are called **arteries**. Blood vessels that carry blood back to your heart are called **veins**. Capillaries are the tiniest blood vessels. They connect arteries and veins.

The biggest artery is the **aorta**. Blood leaves the heart through the aorta. Other arteries branch from the aorta to reach all parts of the body. Then blood flows into tiny veins, which lead to larger veins and back to the heart.

The right side of the heart pumps blood into the lungs. There the blood gets rid of carbon dioxide and picks up oxygen. The oxygen-rich blood flows into the left side of the heart, through the aorta, and to the rest of the body.

LESSON 5 REVIEW Write the answers to these questions on a separate sheet of paper. Use complete sentences.

1) Describe the path of air from your nose to the air sacs in your lungs.

2) What happens in the air sacs?

3) List two purposes of blood.

4) What is blood made of?

5) How do the respiratory system and the circulatory system work together?

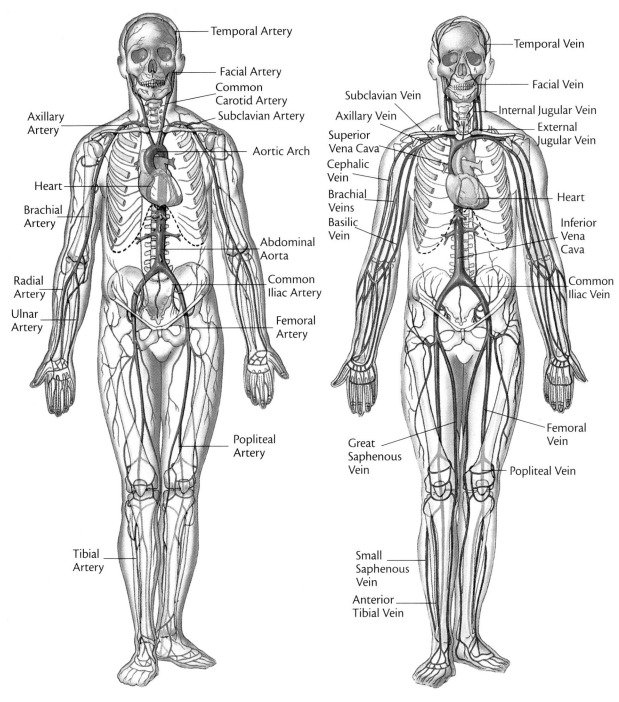

Figure 1.8. The circulatory system

Lesson 6

The Nervous System

Cerebellum
The lower part of the brain, which controls balance and coordination

Cerebrum
The top part of the brain, which controls thinking

Medulla
The part of the brain that connects to the spinal cord

Nervous system
The body system that sends and receives messages throughout the body

Spinal cord
The cable of nerve cells within the bones of the spine

Your **nervous system** allows the parts of your body to communicate with one another. It sends and receives messages throughout your body.

How Does the Nervous System Work?

Your skin, eyes, ears, nose, tongue, and other body parts pick up information about what is happening around and inside your body. This information is sent to your brain through nerve cells. Usually, your brain will receive a message and send a message back, telling that body part what to do. All of this happens very quickly, often without you even knowing about it. Messages to and from the brain are passed from one nerve cell to another at a speed faster than 100 miles an hour!

Your body has billions of nerve cells that reach all parts of your body. These nerve cells form a thick cable called the **spinal cord**. The spinal cord runs through your spinal column to your brain. The nerves in the spinal cord carry messages to and from the brain.

How Does the Brain Control Your Body?

Your brain runs your body. It determines what you think, do, and feel. Your brain has three main parts: the **cerebrum**, the **cerebellum**, and the **medulla**. Figure 1.9 shows these main parts of your brain.

The top part of your brain is the cerebrum. Find the cerebrum in Figure 1.9. The cerebrum has two halves, called hemispheres. Notice that the figure shows only the left hemisphere. The left hemisphere controls the ability to reason, use language, and do math. The right hemisphere controls your artistic ability and your imagination. However, each person's brain is slightly different. Everything you learn, think, and feel is controlled by your cerebrum. Your cerebrum stores your memories. You also use it when you make decisions or link past experiences to new situations. These are only some of the activities of your cerebrum.

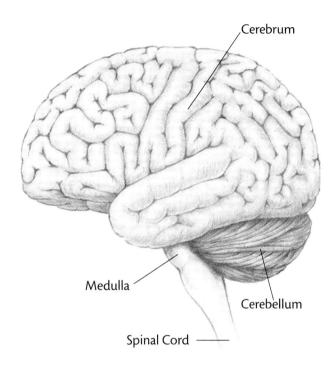

Figure 1.9. The parts of the brain

Below your cerebrum is your cerebellum. The cerebellum coordinates voluntary muscles so that they move smoothly. This part of the brain works closely with the cerebrum. The cerebrum, for example, may decide that you need to lift a box off the floor. It sends a message to the muscles to do the work. The cerebellum also sends messages to the muscles so that all the muscles needed for the job work together smoothly.

The third main part of the brain is the medulla. You can see the medulla on the stem-like part of the brain that connccts to the spinal cord. The medulla is in charge of your involuntary muscles—those that work without your having to think about them. The medulla works day and night to keep your heart beating, your lungs breathing, and your digestive system breaking down food. It also controls swallowing, coughing, and sneezing.

Central nervous system
The brain and the spinal cord

Peripheral nervous system
All the nerves in the body outside the brain and the spinal cord

What Are the Parts of the Nervous System?

Your nervous system is really made of two systems. The brain and spinal cord make up the **central nervous system**. The **peripheral nervous system** includes all the nerves outside the central nervous system. Figure 1.10 shows both of these systems.

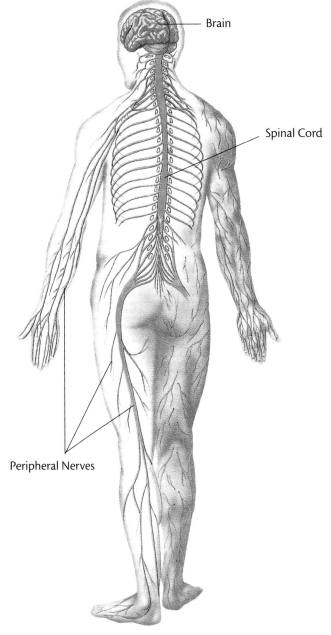

Figure 1.10. The nervous system

42   Chapter 1   The Body Systems

ORDERLY

Health care professionals rely on orderlies. Orderlies assist doctors and nurses in caring for people in hospitals, nursing homes, and clinics. They take care of people's emotional and physical needs. Orderlies check people's health, help keep them clean, and socialize with them. Being an orderly is a good way to try the health care field. Job prospects are excellent. A nursing home association says there is a continued demand for entry-level positions. Some health care facilities offer on-the-job training or pay for classes. Many orderlies like the health field so well, they go back to college or universities to become nurses, therapists, or counselors.

Why would it be more critical to injure the central nervous system rather than the peripheral nervous system?

The main job of the peripheral nervous system is to communicate all the information of the body to the central nervous system. Nerves branch from the spine. These nerves are part of the peripheral nervous system. Other nerves go to all parts of the body, including the arms, the legs, and the organs. All of these nerves are also part of the peripheral nervous system. They send and receive messages to the brain.

LESSON 6 REVIEW Write the answers to these questions on a separate sheet of paper. Use complete sentences.

1) How does the nervous system send and receive messages throughout the body?
2) What are the main parts of the brain?
3) What are the jobs of the cerebrum and the cerebellum? Compare and contrast the jobs.
4) What are the two systems of the nervous system?
5) What is the main job of the peripheral nervous system?

Lesson 7

The Endocrine System

Endocrine system
The body system that uses chemicals to send and receive messages

Like the nervous system, the **endocrine system** controls many activities of your body. It directs activities such as reproduction, growth, and emotions.

How Does the Endocrine System Work?

The endocrine system is made of glands. You can see these glands in Figure 1.11.

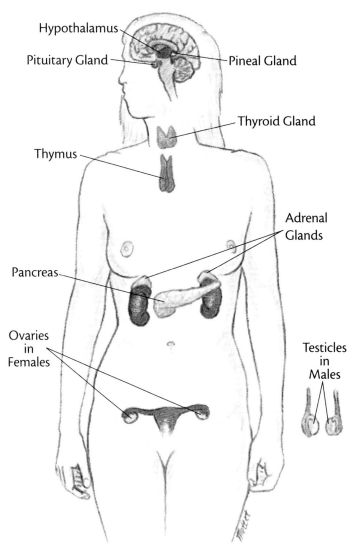

Figure 1.11. The endocrine system

Hormone
A chemical messenger produced by a gland

Secrete
Form and give off

Endocrine glands form and give off, or **secrete**, chemicals called **hormones**. Hormones control activities within the body. They control reproduction, growth, and your emotions.

The hormones are sent directly into the blood. When a hormone is secreted into the blood, it travels to another part of the body. For example, a hormone secreted by a gland near your brain may travel to your reproductive organs. These hormones cause development of your reproductive organs during the teenage years.

Why are hormones secreted only during certain times or during specific phases of life?

DIABETES

Diabetes is a disease in which a person's body does not make enough of the hormone insulin. As a result, the blood carries an unsafe amount of sugar. Infections and poor blood flow once were common side effects. Until the early 1900s, most people with diabetes did not live long.

In 1921, Canadian doctor Frederick Banting began to research diabetes. The following year, he perfected an insulin injection for diabetes patients. Since then, millions of people with diabetes have been able to live longer by taking insulin or a pill to stimulate the pancreas to produce insulin. Today, researchers are working on a way to save a person's blood from their umbilical cord at birth to use to grow cells to help stimulate the pancreas or replace diseased tissue. Perhaps in the future, insulin injections will not be needed.

Adrenal gland
The endocrine gland that releases several hormones

Adrenaline
The hormone that increases certain body functions

Pituitary gland
The endocrine gland attached to the base of the brain

Thyroid gland
The endocrine gland that affects a person's energy

What Do the Pituitary, Thyroid, and Adrenal Glands Do?

The **pituitary gland** has a very important role in the body. It secretes many different kinds of hormones and is sometimes called the body's master gland.

The pituitary gland is a pea-sized gland attached to the base of the brain. It releases hormones that help a child develop into a young adult. These hormones cause the sex organs to mature. The pituitary gland also produces growth hormones that affect a person's size.

The **thyroid gland** is a large endocrine gland. It produces hormones that help the body change food into energy. Too much of these hormones can cause a person to feel overly energetic. Too little of it causes a person to feel tired. The pituitary gland secretes a hormone that causes the thyroid to act.

An **adrenal gland** is on the top of each kidney. These glands secrete hormones that help the body maintain the proper water balance and cope with stress. They also secret a hormone called **adrenaline**. Adrenaline causes your heart to beat faster in an emergency.

LESSON 7 REVIEW Write the answers to these questions on a separate sheet of paper. Use complete sentences.

1) What are endocrine glands?

2) What are hormones?

3) How does the endocrine system control development of your reproductive organs?

4) Why is the pituitary gland sometimes called the master gland?

5) How does adrenaline affect the body in an emergency?

Lesson 8

The Reproductive System

Ovary
The female organ that stores eggs

Ovulation
The monthly process of releasing an egg

Puberty
The period when children develop into adults and reach sexual maturity

Reproductive system
The body system responsible for making a baby

Reproduction is the process through which a male and a female produce a child. The ability to reproduce is one of the most amazing features of the human body. It is carried out by the **reproductive system**. It is the only body system that differs between males and females.

What Are the Parts of the Female Reproductive System?

The female reproductive system is shown in Figure 1.12. Females are born with more than a million eggs. The eggs are stored in the **ovaries**. The eggs begin to mature when the pituitary gland releases hormones that act on the ovaries. The mature eggs are then released in a monthly process known as **ovulation**. This happens during **puberty**. Puberty is the time when children begin adult development and reach sexual maturity.

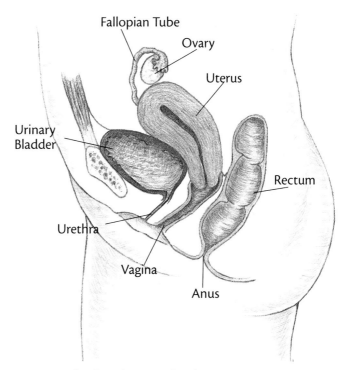

Figure 1.12. The female reproductive system

Menstruation
The monthly flow of an egg and extra tissue from the uterus

Sperm
The male sex cells

Testis
The male organ that makes sperm

Uterus
The female organ that holds a growing baby

Vagina
The birth canal through which a baby is born

What Happens When an Egg Is Released?

Once released, an egg moves through a tube to the **uterus**. On its way to the uterus, the egg may be joined by a **sperm**, or male sex cell. If the egg is joined by a sperm, it attaches itself to the wall of the uterus.

The uterus will become larger to make room for the egg and sperm. The egg and sperm will make new cells. The uterus will also provide blood, food, and oxygen to the developing baby.

If the egg is not joined by a sperm, it leaves the body. This monthly passing of an egg and other tissue is called **menstruation**. Menstruation is the blood flow that takes place each month in a woman's body.

How Is a Baby Born?

A baby will grow in the uterus until it is time for it to be born. Then the uterus opening will expand to allow the baby to move out of the uterus.

The **vagina** is the birth canal. This tube leads from the uterus to the outer sexual organs. It is the path the baby takes when it is born.

What Are the Parts of the Male Reproductive System?

In males, the sex cells are called sperm. Sperm are made in the male sex organs called **testes**. The testes begin making sperm when the pituitary gland sends hormones to act on the testes during puberty. Figure 1.13 shows the male reproductive system.

The testes are in a sac outside the body. The temperature of the sac is about five degrees lower than the temperature of the rest of the body. Sperm do not live long at the body's normal temperature.

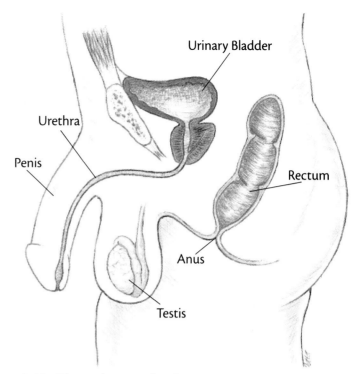

Figure 1.13. The male reproductive system

Penis
The male organ used to deliver sperm and to urinate

Sexual intercourse
Inserting the penis into the vagina

How Are Sperm Released?

The sperm are stored in coiled tubes on the side of the testis. Mature sperm flow into the urethra. The urethra leads to the outside of the body through the penis. Both sperm and urine exit the body through the urethra, but not at the same time.

How Does New Life Begin?

Once both a male and female reach puberty, it is possible for them to reproduce a new life. **Sexual intercourse** is nature's way of joining a male sperm and a female egg to form a new life.

What reasons might explain why a sperm and an egg are sometimes unable to join?

During intercourse, sperm pass through tubes to the **penis**. The penis is the male reproductive organ. It has many small blood vessels. These vessels fill with blood, making the penis firm. The sperm travel through the penis to enter the female body through the vagina. During intercourse, millions of sperm are released into the female. It takes only one sperm cell to join with an egg. If this happens, the growth of a new baby begins.

The Body Systems Chapter 1

Action for Health

SELF-EXAM FOR EARLY SIGNS OF CANCER

Learn to examine yourself for changes that may be early warning signs of cancer.

For Females: Self-Exam of the Breasts

Females should do a self-examination after each menstrual period. The three ways to examine breasts are before a mirror, in the shower, or lying down.

Stand before a mirror and look at your breasts for any unusual depressions or lumps. Look for unusual changes.

In the shower or lying down, use the left hand to examine the right breast and the right hand to examine the left breast. Use your first three fingers. Examine each breast in a circle around the nipple until you cover the entire breast. Also, check the nipple and the area between the armpit and the breast.

For Males: Self-Exam of the Testes

Cancer of the testes is one of the most common cancers in males between the ages of 20 and 35. Most testicular cancers are first discovered during self-examination. Examine the testicles once a month after bathing. Gently roll each testicle between the thumb and fingers, noting any abnormal lumps.

LESSON 8 REVIEW Write the answers to these questions on a separate sheet of paper. Use complete sentences.

1) Compare how eggs are produced with the way sperm are produced.
2) What is the difference between ovulation and menstruation?
3) Describe the path of an egg through the female reproductive system.
4) Describe the path of sperm out of the male's body.
5) How do the male and female reproductive systems work together to reproduce?

Chapter Summary

■ Tiny units, called cells, make up all living things, including the human body.

■ Cells form tissues, and different types of tissues form organs. Groups of organs form a body system.

■ The skin is the largest organ of the body. It has three layers that cover and protect your body.

■ The skeletal system is made up of 206 bones. Bones give your body its frame and protect underlying organs.

■ The muscular system allows your body to move. You move when a muscle pulls a bone.

■ Muscles that you control are called voluntary muscles. Muscles you do not control are called involuntary muscles.

■ The digestive system breaks food down into a form that your cells can use. Blood carries the digested food to the cells.

■ Food that is not digested passes out of the body as solid waste.

■ Other wastes pass out of your body in a liquid form called urine. Urine is formed in the kidneys and leaves the body through the urethra.

■ Your respiratory system brings fresh air into your lungs. The circulatory system carries oxygen and carbon dioxide to and from your body cells.

■ The nervous system allows the different parts of your body to communicate with one another. The different parts of your body send information to your brain and spinal cord.

■ Glands in the endocrine system secrete chemicals called hormones. Hormones travel in the blood to other parts of the body.

■ The male and female reproductive systems work together to produce a baby.

Chapter 1 Review

Comprehension: Identifying Facts

On a separate sheet of paper, write the correct word or words from the Word Bank to complete each sentence.

WORD BANK		
blood	endocrine	kidneys
bones	epidermis	muscle
cerebrum	glands	spinal cord
cells	hormone	uterus
digestive		

1) The pituitary, thyroid, and adrenal glands are part of the _____ system.
2) The _____ is made up of cells and plasma.
3) The skeletal system is made up of 206 _____.
4) The _____ system breaks food down into a form cells can use.
5) The brain and the _____ make up the central nervous system.
6) A _____ is a chemical produced by a gland.
7) Endocrine _____ send hormones through your bloodstream.
8) The largest part of the brain is the _____.
9) A tissue is made up of many _____ working together to do the same job.
10) When you move, a _____ is pulling a bone.

11) The three layers of the skin are the _____, the dermis, and the subcutaneous layer.

12) Urine is formed by the _____.

13) After a sperm joins an egg, the egg attaches itself to the wall of the _____.

Comprehension: Understanding Main Ideas

Write the answers to these questions on a separate sheet of paper. Use complete sentences.

14) How does your skin protect your body?

15) How do the respiratory and circulatory systems work together to bring oxygen to your cells?

16) What are three purposes of your blood?

17) How is the endocrine system different from the nervous system?

18) How do the muscular and skeletal systems work together so that you can move?

Critical Thinking: Write Your Opinion

19) Which body system do you think is more important than the others? Explain your answer.

20) Many doctors specialize in treating one specific body system. Suppose you were a doctor. Which body system would you specialize in? Why?

Test Taking Tip Sometimes it is easier to learn vocabulary words if you break them into their word parts.

Chapter 2

Hygiene and Fitness

Taking care of yourself can be one of the most important things you do for yourself. Caring for your skin, hair, nails, teeth, eyes, and ears can help keep you healthy. Exercising and getting enough rest also are important for your health. Taking care of yourself can help you look and feel better. When you feel good about yourself, other people usually feel good about you, too.

In this chapter, you will learn about ways to take care of yourself. You will learn how keeping clean and taking care of your body are important. You will also learn how exercise and rest can help you. Finally, you will learn how taking care of yourself can help your physical, social, and emotional health.

Goals for Learning

▶ To explain the purpose of basic hygiene
▶ To describe ways to protect skin, hair, nails, teeth, eyes, and ears
▶ To define cardiovascular fitness and explain its importance
▶ To identify the benefits of regular exercise and the parts of an exercise program
▶ To explain why the body needs rest and sleep

Hygiene

Acne
Clogged skin pores that causes pimples or blackheads

Practicing good hygiene is important for your health. Hygiene refers to things that you do to promote your health. Caring for your skin, hair, nails, teeth, eyes, and ears is a part of hygiene.

How Can You Take Care of Your Skin?

Your skin covers and protects your body. Keeping your skin clean is part of good skin care. Washing with soap and water is the best way to keep your skin clean. This can help remove dirt and germs from the outside of your skin.

What are some reasons why one person might need to bathe more often than another person?

Washing with soap and water also helps reduce body odor. Perspiration has no odor. Odor results when perspiration contacts germs on your skin. How often you bathe depends on how active you are and what skin type you have. People who are very active or who have oily skin need to bathe more often.

Strong winds, cold, and sun dry the skin's outer layer. Covering your face and hands in cold or windy weather helps protect your skin. Too much sun can cause sunburn, wrinkled skin, and skin cancer. Wearing sunscreen helps protect your skin. Sunscreens are rated by a sun protection factor (SPF). An SPF of 15 or higher gives the best protection.

Nutrition Tip

Eat plenty of fruits and vegetables. They can provide vitamins and minerals your skin needs to stay healthy.

Skin Problems

A common skin problem for teenagers is **acne**. Acne is clogged skin pores, or openings in the skin. Teenagers have a normal increase in hormones. Hormones cause oil glands to produce more oil. If the oil plugs your pores, you may get blackheads or pimples. A blackhead is an oil plug that gets dark when air contacts it. A pimple forms when the skin around the blackhead becomes red and filled with pus. Here are some things you can do to help control acne.

- Wash right after exercise to clean away sweat and germs that can clog pores. Use a clean washcloth every day.
- Shampoo your hair often to limit acne on your forehead, neck, and shoulders.

technology

LASER SURGERY

Science fiction movies show lasers as a weapon used against the enemy. Today lasers are often used for surgery. Laser surgery was first used in the mid-1960s. A laser produces a strong, narrow beam of light. It can be controlled accurately. Lasers can seal blood vessels. They can cut through skin and remove scars and wrinkles. They can change the shape of part of the eye. Some lasers can help nearsighted people see without glasses or contacts. Lasers can help people who snore. Dentists can even use lasers to whiten teeth.

Caries
Cavities in the teeth

Dermatologist
A doctor who takes care of skin

Plaque
A layer of bacteria on teeth

- Eat a well-balanced diet. Get plenty of rest and exercise.
- Do not squeeze or pick pimples and blackheads.

If you have a skin problem, you might visit a **dermatologist**. A dermatologist is a doctor who takes care of skin problems.

How Can You Care for Your Hair and Nails?

Shampooing your hair often helps keep it clean. Brushing your hair distributes natural oils and makes your hair shiny. Too much heat can damage your hair. If you use a hair dryer or curling iron, use a low heat setting.

Nails protect the ends of your fingers and toes. Scrubbing your nails will get rid of dirt and germs under them. Cutting your nails evenly and filing rough edges will help protect them.

How Can You Take Care of Your Teeth?

Brushing and flossing your teeth at least once a day can help remove **plaque**. Plaque is a sticky, colorless layer of bacteria that forms on teeth. Plaque causes **caries**, or cavities, in the teeth. Dental caries is a leading cause of tooth loss in children. Plaque also causes gum disease, a leading cause of tooth loss in adults.

Decibel
A unit that measures sound

Optometrist
A specialist in eye examinations and corrective lenses

Orthodontist
A dentist who treats crooked or crowded teeth

Making regular visits to the dentist is important. A dentist can check your teeth for cavities or other dental problems. Visits to the dentist also include a thorough cleaning of the teeth.

Many people have teeth that are crowded or crooked. Teeth that are out of position are harder to clean. This makes tooth decay and gum disease more likely. An **orthodontist** is a dentist who treats crowded or crooked teeth. Treatment can include braces or wires to move the teeth into the proper position.

How Can You Care for Your Eyes and Sight?

You can help protect your eyes from infection, disease, and injury. Wearing safety glasses, goggles, or a helmet can protect your eyes from damaging light, dirt, or sports injuries. Wearing glasses or contact lenses helps you see better and prevents eyestrain. If you wear contact lenses, it is important to follow instructions for cleaning and wearing them.

Your vision is normal if you see a clear image. If you do not see clearly, an **optometrist** can prescribe glasses or contact lenses. An optometrist is a specialist in eye examinations and corrective lenses. Regular checkups with an optometrist can

PROTECT YOUR HEARING

The loudness of sound is measured in **decibels** (db). Here are some decibel levels of some common sounds:

Normal breathing	10 db
Whisper	30 db
Normal conversation	60 db
Truck traffic	90 db
Rock concert	120 db
Jet engine at takeoff	140 db

You can help protect your hearing from harmful noise. Keep the volume low on radios, CD players, and TVs, especially when wearing headphones. Get away from any sound that is too loud. Decibel levels higher than 85 can cause hearing loss. In jobs where noise is greater than 85 db, workers must wear earplugs or earmuffs to protect their hearing.

If you wear contacts, be sure to follow instructions for cleaning them.

Ophthalmologist
A doctor who specializes in diseases of the eye

help correct vision problems and prevent disease. An **ophthalmologist** is a doctor who specializes in eye diseases.

How Can You Care for Your Hearing?

A common reason for hearing loss is loud noise. Infection, allergy, buildup of wax, and injury can also cause hearing loss. If you have problems with your hearing, see a doctor.

There are things you can do to help protect your ears and your hearing. Wear a head guard when playing contact sports. Be careful to swim only in clean water. Clean only the outside of your ears. Never put a swab or other object into your ear. Stay away from loud noises as much as you can.

LESSON 1 REVIEW Write the answers to these questions on a separate sheet of paper. Use complete sentences.

1) What are two ways you can take care of your skin?
2) How can you remove plaque from your teeth?
3) How can you protect your eyes?
4) How can you protect your ears and hearing?
5) What are some noisy places where you think wearing earplugs might be a good idea?

Healthy Subjects

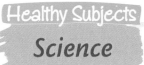

NOISE POLLUTION

Noise pollution is a health problem. Loud noise can cause hearing loss. Many industries and companies have worked to reduce loud noise. They use special materials in walls and ceilings to reduce sound levels. They place noisy machines in cases that cut down noise. Airport workers wear earplugs or earmuffs when jets take off. Other people who work in noisy places also wear ear protectors.

Lesson 2

Fitness

Aerobic exercise
An exercise that raises the heart rate

Blood pressure
The force of the blood against the walls of the arteries when the heart beats

Cardiovascular fitness
The condition of the heart, lungs, and blood vessels

Heart rate
The number of times the heart pumps blood each minute

Physical fitness
The body's ability to work, exercise, and play without tiring

Physical fitness is your body's ability to exercise, work, and play without tiring easily. If you are physically fit, you have enough energy to do all the things you want to do. Physical fitness is an important part of your good health. It affects your emotional, social, and physical well-being.

How Is Exercise Important to Your Health?

Cardiovascular fitness is the condition of the heart, lungs, and blood vessels. You can improve your cardiovascular fitness by exercising. Heart disease is the leading cause of death in this country. Heart disease often begins in childhood. Exercising to improve cardiovascular fitness can help reduce the risk of getting heart disease.

Your heart is a muscle. When you exercise, more blood enters the heart. The heart works harder to push more blood into the blood vessels. Then your heart muscle gets stronger.

Your heart pumps blood containing oxygen through your body. Your arteries are blood vessels that carry blood away from the heart to the rest of your body. Your veins carry blood from your body to your heart. Exercising helps blood flow through your body and back to your heart.

How well your heart and blood vessels are moving blood through your body can be measured. Your **heart rate**, or pulse rate, is the number of times your heart pumps blood each minute. **Blood pressure** is the force of blood against the walls of the arteries when the heart beats. High blood pressure is a sign that the heart and blood vessels have to work too hard. High blood pressure can lead to stroke, heart attacks, and kidney failure.

Aerobic exercises can help improve cardiovascular fitness. These are exercises that raise your heart rate. When you do aerobic exercises, your body takes in more oxygen. Some kinds of aerobic exercises are walking, running or jogging, biking, swimming, and cross-country skiing.

Fitness Tip

Wear washable clothes when you exercise. Make sure the clothes are comfortable and loose. Clothes made of cotton allow your body to stay cool. Wearing proper shoes for physical activities can help prevent injuries.

How Can You Exercise Properly?

No matter what kind of exercise you do, your workout should include a warm-up, an exercise period, and a cool-down. Skipping the warm-up and cool-down can be harmful to your body.

Warming up increases the blood flow to your muscles and prepares your body for exercising. Warming up allows your heart rate to increase gradually. A warm-up should take about five minutes. Walking and then stretching is a good way to warm up.

The exercise period should include exercises that make your heart and lungs work hard. It may include exercises such as sit-ups and push-ups to improve muscle strength. You can do stretching and bending exercises as part of the exercise period. These exercises help the body twist and turn easily. They help the body move and change positions smoothly.

Every workout should end with a cool-down. If you stop exercising suddenly, you can become dizzy. To cool down, continue to exercise gently at a slower pace. Do some stretching exercises. The cool-down should take about five minutes.

Who Benefits From Exercise?

People who have a risk of heart disease can be greatly helped by exercise. In fact, everyone can benefit from some form of exercise. Regular exercise is a healthy habit that can have a positive effect on your physical, social, and emotional health. The top calorie burning exercises to help you be physically fit are found in Appendix A.

Regular exercise can improve your health. Exercise gives you more energy and helps you feel better.

Careers

FITNESS INSTRUCTOR

Do you like to stay fit? If so, you might like to become a fitness instructor. Many companies have health and wellness programs. These programs can help workers develop healthy habits. Fitness instructors work with companies to help plan these programs. They teach workers how exercise helps the body. They help workers make exercise plans. They also show workers the correct way to exercise. Instructors may work with doctors and therapists to help workers with medical problems. You need special certification classes to become a fitness instructor. You need to study about the parts of the body and how they work together.

Physical Health

Exercise firms and strengthens your muscles and helps your body move more smoothly. It also helps your heart and lungs to work better. When you exercise, you have more energy, you feel better, and you look better

Regular exercise helps you keep in shape at a weight that is right for you. Exercise burns calories. It also helps the body continue to burn calories at a faster rate.

What might you say to a person who does not exercise because he or she thinks it is boring?

Social Health

Exercise is fun. It helps you enjoy your extra time. Exercising is a good way to make new friends. Walking, jogging, skiing, bicycling, swimming, and many other sports are good exercises and good activities to do with friends.

Emotional Health

Exercise helps reduce the symptoms of stress. It helps you take your mind off your problems. After exercising, you feel refreshed and relaxed.

Why Are Rest and Sleep Important to Fitness?

When you rest and sleep, your heart rate and breathing slow down. Your blood pressure drops and your muscles relax. Your body uses less energy.

Writing About Health

Think about a physical activity that is fun for you. Write about when you do the activity. Tell why you enjoy it.

When you are tired, you may have a hard time paying attention and remembering things. You may make more mistakes than you usually do. You are more likely to get ill or injure yourself when you are tired. You might find that you are less able to cope with upsetting situations.

Most young people need about eight or nine hours of sleep each night. If you are active, you may need extra rest. Relaxing is a way to rest during the day.

LESSON 2 REVIEW Write the answers to these questions on a separate sheet of paper. Use complete sentences.

1) What two things measure how well the heart is moving blood through the body?
2) How does exercise improve your cardiovascular fitness?
3) What are three important parts of a workout?
4) Name three parts of overall health that can be improved by exercise.
5) How might lack of sleep affect how well you study?

WOMEN IN SPORTS

Around 1900, women in America were not allowed to play most sports. They did play croquet and archery. Women had to wear long dresses with long sleeves when they played. These clothes made it hard for women to play.

Over the years, women started playing volleyball, baseball, and basketball. Today, women play as many sports as men do. They have become champion figure skaters, runners, and tennis players. They have succeeded in gymnastics, water sports, and many other sports.

Chapter Summary

- Caring for skin, hair, nails, teeth, eyes, and ears is a part of hygiene.

- One important way to care for the skin is to wash with soap and water.

- Protecting skin from strong winds, cold, and sun is important.

- Acne can be controlled by washing often and avoiding squeezing pimples and blackheads.

- Shampooing hair often helps keep it clean.

- Keeping nails clean will get rid of dirt and germs under the nails.

- Brushing and flossing teeth helps remove plaque.

- People can help protect their eyes and their sight by wearing safety equipment as needed and having regular checkups.

- People can protect their ears and hearing by wearing head guards for playing sports and avoiding loud noise as much as possible.

- People can improve cardiovascular fitness by exercising.

- How well the heart and blood vessels move blood through the body can be measured by heart rate and blood pressure.

- Aerobic exercises help improve cardiovascular fitness.

- A good exercise program includes a warm-up, an exercise period, and a cool-down.

- An exercise period can include exercises to improve muscle strength, exercises to make the body move more smoothly, and stretching exercises.

- Everyone can benefit from some form of exercise.

- Exercise has a positive effect on physical, social, and emotional health.

- A person who is tired may experience difficulty paying attention and other difficulties.

- Most teenagers need about eight or nine hours of sleep each night.

Chapter 2 Review

Comprehension: Identifying Facts

On a separate sheet of paper, write the correct word or words from the Word Bank to complete each sentence.

WORD BANK	
acne	gum disease
aerobic exercises	heart rate
blood pressure	ophthalmologist
cardiovascular fitness	optometrist
caries	orthodontist
crooked teeth	physical fitness
dermatologist	plaque

1) The ability to work, exercise, and play without tiring is called _____.

2) An _____ is a doctor who specializes in diseases of the eye.

3) _____ can be treated with braces or headgear.

4) The number of times the heart pumps blood each minute is called the _____.

5) Brushing and flossing can help remove _____ from the teeth.

6) A doctor who takes care of skin is called a _____.

7) _____ is the leading cause of tooth loss in adults.

8) A dentist who treats crooked or crowded teeth is called an _____.

9) Exercises that improve _____ can help reduce the risk of heart disease.

10) The leading cause of tooth loss in children is dental _____.

11) High _____ is a sign that the heart and blood vessels have to work too hard.

12) An _____ can give eye examinations and prescribe glasses or contact lenses.

13) _____ are exercises that raise the heart rate.

14) _____ is a common skin problem among teenagers.

Comprehension: Understanding Main Ideas

Write the answers to these questions on a separate sheet of paper. Use complete sentences.

15) Why should a person avoid squeezing pimples or blackheads?

16) How can you remove plaque from your teeth?

17) How does exercise help your heart?

18) What are some kinds of aerobic exercise?

Critical Thinking: Write Your Opinion

19) Why do you think it might be a good idea to do exercises other than those that are part of team sports?

20) What advice might you give to a friend who uses headphones to listen to loud music?

Test Taking Tip If you know you will have to define certain terms on a test, write the term on one side of a card. Write its definition on the other side. Use the cards to test yourself or work with a partner to test each other.

Chapter 3

The Family

Most people belong to a family and depend on the support they gain from their family. There are many different types of families. However, all families provide similar benefits for their family members. Families help their members act responsibly, both as a group and as individuals.

In this chapter, you will learn about the family life cycle and how new families are formed through marriage. You will learn about parenting and the responsibilities that parents have. You will also learn about different types of families and what makes a family healthy. Finally, you will learn about problems in families and where to get help with the problems.

Goals for Learning

▶ To explain the place of marriage in the family life cycle
▶ To identify some characteristics of a healthy marriage
▶ To describe parenting and responsibilities parents have
▶ To describe some characteristics of a healthy family
▶ To identify where families can get help for problems

Lesson 1

The Family Life Cycle

Commitment
Love, dedication; a pledge of trust

Extended family
A family who includes many people from different generations

Family life cycle
The changes in a family over time

Families change, just as individuals do. The changes in families over time are known as the **family life cycle**. A family life cycle includes events such as marriage and the birth of children.

Each stage in the family life cycle presents the family with new challenges and tasks. Families who meet these challenges lay the foundation for health and strength in the family.

What Is a Family?

Families come in many forms. Some families are small and some are large. Family members can include many people, such as parents, children, grandparents, aunts, uncles, brothers, and sisters. Families who include many people from different generations are called **extended families**.

A new family is formed when two people marry. This is the first stage in the family life cycle. The newly married couple must adjust to each other and the demands of married life. The couple will form new relationships with each other's extended family and friends. The couple will also share the duties of the family. Every family will solve these challenges in its own way. Discussion and **commitment** are important keys to making the relationship work.

What Happens When the Family Grows?

Once a married couple has adjusted to the challenges of marriage, many new changes can occur. Children may be added to the family group. Family members and their roles change as the family changes.

When a child is born into the family, the married couple has new responsibilities as parents. This is the second stage in the family life cycle. The parents must meet the needs of the new child while staying healthy themselves.

Discipline
Correct behavior

One of the challenges of this stage is understanding the new child's needs. Infants need love and attention, as well as food, clothing, and shelter. As the child grows, the parents need to set limits for the child and teach **discipline**. It is also important for parents to sometimes see "through the eyes of a child" and not get angry or impatient at normal mistakes that the child makes.

What Are Characteristics of Healthy Marriages?

There is no single recipe for a successful marriage, but the following are some things that have been identified as important.

- Agreeing on money matters
- Having similar interests
- Knowing each other well before marriage
- Accepting and supporting each other
- Agreeing about having children and about how to discipline children
- Having common goals
- Sharing household tasks
- Having similar family backgrounds and good relationships

Parents need to teach limits and understand their children's needs.

68   Chapter 3   *The Family*

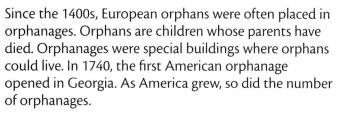

ORPHANAGES AND ADOPTION

Since the 1400s, European orphans were often placed in orphanages. Orphans are children whose parents have died. Orphanages were special buildings where orphans could live. In 1740, the first American orphanage opened in Georgia. As America grew, so did the number of orphanages.

In the 1900s, adoptions of children into families began to replace orphanages. Often, orphans live with a foster family while waiting for adoption. U.S. adoption figures show that the number of adopted children has increased since the 1960s. About 50,000 American children where adopted during the 1990s. In 1997, there were 13,620 adoptions of children from foreign countries.

Adolescent
A child between the ages of 13 and 17

When children become teenagers, they begin to prepare to live an independent life. This is the third stage in the family life cycle. Parents will continue to set rules for the safety of their **adolescents**. Parents also allow them to become independent and form close relationships outside the family.

This stage of the life cycle is often a time when both adolescents and parents struggle for the right balance between too many and too few rules. Sometimes adolescents want more freedom but are not ready to take the responsibility that goes along with it. Communication, patience, and caring are some skills that families can use while they try to address the challenges of this stage.

Health Tip

Talk about and work out problems for healthy relationships.

How Does Aging Affect the Family?

As family members age, their roles and relationships within the family change. Some may leave and form families of their own. Older family members may die, and the family will adjust and cope with their loss.

When they are old enough, children leave home and begin leading their own lives. This is the fourth stage in the family life cycle. The family returns to being a couple again for the second time. Parents often feel a sense of loss when their

The Family Chapter 3 **69**

children leave home. Sometimes they even feel they have lost their importance in the family.

Married couples may change and renew their relationship to each other. They may also develop new interests or hobbies to keep themselves occupied. Eventually, the parents will form new relationships with their children's friends, spouses, and in-laws. Parents may also learn new roles as grandparents.

In the fifth and final stage of the family life cycle, a couple must learn to accept growing limitations and disabilities. Sometimes, couples must adjust to a new home or community. At this stage, a family member may learn to cope with the death of the spouse. Older family members continue to contribute to the family by sharing their wisdom, knowledge, and family history.

LESSON 1 REVIEW Write the answers to these questions on a separate sheet of paper. Use complete sentences.

1) What is an extended family?

2) What are some of the challenges in the second stage of the family life cycle?

3) What are some characteristics of a healthy marriage?

4) What are the five stages of the family life cycle? Briefly describe each stage.

5) How do older relatives contribute to a family?

FINDING CARE FOR FAMILY MEMBERS

Caring for the elderly is an important issue in the United States. The number of elderly people has grown steadily over the years, and will continue to grow as the baby boom generation ages. Most people are familiar with day care programs for children, which provide care and education while parents work. Similar programs are available for older relatives as well. Investigate which facilities are available in your community for both younger and older family members. Compare their similarities and differences. If you were to choose a facility for a child or grandparent, what qualities would you look for?

Lesson 2

Dealing With Family Problems

Depressed
Extremely sad

Divorce
The end of a marriage

Separation
A period when a married couple stops living together

When families successfully deal with changes and challenges, they lay the foundation for a healthy family unit. However, some changes, such as divorce, illness, and death, may be too difficult for families to deal with on their own. Troubled families can get help to survive and recover from such problems.

How Does Stress Affect a Family?

Some of the stresses that families deal with are a normal part of everyday life. Family members work the problems out by changing the way they relate to one another. For example, the birth of a new child or relocation of an older family member may require people to take on new roles and responsibilities.

Unexpected events can increase family stress and cause problems. If a family member loses a job, the family must change to deal with lost income. This may cause a physical and emotional burden for all family members. Tragedy, such as an illness, injury, or the death of a family member, is also a stressful change. The loss or disability of a family member who has a strong, supportive role in the family can be very difficult. Other family members may feel they have less emotional support and no one to turn to.

What are some special challenges for children in single-parent households?

How Do Separation and Divorce Affect a Family?

Sometimes a couple decides that they cannot live together anymore. They may agree to a **separation** so that they can think about how to make their marriage work. If they believe that the marriage is not healthy and that they cannot improve it, they may decide to **divorce**.

Separation and divorce can be difficult for children. They may worry or feel **depressed**. They may feel that they are losing a parent. Children may even believe that they have caused the divorce.

The Family Chapter 3

Violent
Doing actions or words that hurt people

Fitness Tip

Go for a walk or a run when you feel unhappy.

How Do Families Deal With Problems?

People often react to serious problems in several stages: denial, anger, adjustment, and acceptance. *Denial* is a common response when people are not ready to cope with a problem. However, denial does nothing to solve the problem. Feelings of *anger,* sadness, or loneliness often overcome the denial. If high stress levels continue, family members may argue and ignore one another's needs and feelings. The *adjustment* stage happens when the family recognizes the problem and begins working on a solution. Healthy family solutions include setting goals for change and being willing to compromise. *Acceptance* is reached when a family makes successful attempts to adjust. Family members must learn to accept that some events are beyond their control. However, they can work to change those things that can be changed and to move on.

How Do Violence and Abuse Affect a Family?

Most families find ways to deal with stress and problems. However, problems in some families are so severe that the family becomes unhealthy. Family members no longer support one another and may even become **violent**.

Family members can help each other.

FAMILY COUNSELOR

Many families need help solving their problems. Some families today face problems such as alcohol and drug abuse, depression, and violence. Some families may just need help adjusting to normal stresses and changes. Family counselors help families with these problems. Crisis intervention may be necessary if there is substance or physical abuse. Family counselors must be able to understand how all the members of the family feel. They must be able to help the family members form trusting, caring relationships with one another. Requirements vary from on-the-job experience to college level programs in social work or psychology. Never before has the need for family counselors been so great.

Child abuse
An action that harms a child

Emotional abuse
A mistreatment through words, gestures, or lack of affection

Sexual abuse
Any sexual contact that is forced on a person

Violence may be caused by anger, stress, drugs, or alcohol. Sometimes people learn to be violent because that was how they were raised. **Child abuse** is a form of violence. Victims of physical abuse may suffer bruises, broken bones, or even death. Harsh words, threats, and lack of affection are forms of **emotional abuse**. Victims of emotional abuse are often depressed and scared. They may avoid contact with others. **Sexual abuse** is any sexual contact that is forced on someone.

Every state has laws to protect children from abuse. These laws require people who work with children to notify local authorities if abuse is suspected. Victims may be afraid to report abuse because they feel shame or fear. Sometimes children may feel responsible for the abuse. However, in *no* case are victims responsible for abuse. Telling a trusted adult is very important. It is the first step in getting help. With help, victims of abuse are able to overcome the pain. Without help, victims sometimes become abusers themselves.

Writing About Health

You have read about some of the stresses that can cause problems in families. Write about the effects of one kind of stress on a family. Tell how the family might respond to maintain a healthy family.

Where Can Family Members Go for Help?

Sometimes a family needs help to solve a problem. Admitting that a problem exists and asking for help is the most important step. Denying the problem or failing to address it can cause greater stress later.

Stressful events occur in all families. Asking for help and working together to solve problems strengthens the family. Most communities have support groups to help families deal with serious problems.

Caring adults or friends are good sources of support. Sometimes a parent, sibling, or other relative can help a family through difficult times. In school, some staff members are trained to assist students as they work through problems. When family problems become too difficult to handle, professional counselors, social workers, and psychologists can help the family.

It is helpful to keep a positive attitude while working on solving family problems. Talking with a caring friend or adult or writing feelings down in a diary can help avoid depression. This can give family members more confidence as they concentrate on solving a problem.

LESSON 2 REVIEW Write the answers to these questions on a separate sheet of paper. Use complete sentences.

1) What are some events that cause stress in families?
2) Who can offer support to you about your problems or stress?
3) What are the stages for reacting to serious problems?
4) What should a person do if he or she is being abused?
5) How can families respond positively to stressful situations?

Chapter Summary

- A new family life cycle begins with marriage. Each stage in the family life cycle presents the family with new challenges.

- Parents are responsible for the health, safety, and well-being of their children.

- Parents need to learn certain skills to help their children grow physically and emotionally. They must set rules and provide discipline.

- Similar family backgrounds and having common interests and goals contribute to a healthy marriage.

- The family life cycle is always growing and changing.

- In the fifth stage of the family life cycle, a family deals with the aging of older family members.

- Separation, divorce, aging, illness, and death are changes that affect families. To be successful, family members support each other as they adjust to these changes.

- Normal stresses as well as unexpected events occur in all families. In most cases, family members work to solve these problems together.

- Families can seek help for serious problems. Caring friends, professional counselors, and trained school staff offer support and help.

- Family members often go through several stages when they react to problems: denial, anger, adjustment, and acceptance.

- A family in which violence and abuse occur is unhealthy. Violence and abuse are never acceptable.

Chapter 3 Review

Comprehension: Identifying Facts
On a separate sheet of paper, write the correct word or words from the Word Bank to complete each sentence.

WORD BANK		
adoption	divorce	roles
challenges	elderly	rules
cope	family life cycle	sexual abuse
counselor	independence	violence
depression	responsibilities	

1) A new _____ begins when two people marry.
2) Each stage in the family life cycle presents the family with _____.
3) A married couple must share family _____.
4) Parents need to set _____ for children to keep them safe.
5) When children grow into adolescents, they begin to prepare for _____.
6) When older family members die, the family must _____ with their loss.
7) _____ has replaced most orphanages.
8) It is important to find ways to care for the _____ as they grow older.
9) When families change to meet challenges, family members may take on new _____.
10) When you experience a loss, you may feel _____, or a feeling of deep sadness.

11) Although it is not true, children sometimes feel they are the cause of a _____.

12) A professional _____ can help a family work through difficult problems.

13) Child abuse is a form of _____ directed at children.

14) Victims of _____ are never responsible for the other person's behavior.

Comprehension: Understanding Main Ideas

Write the answers to these questions on a separate sheet of paper. Use complete sentences.

15) What are some ways families address the challenges of their changing lives?

16) Why do people sometimes react by denying a difficult problem?

17) Give an example of emotional abuse and describe how it might affect the victim.

18) List five characteristics of a healthy marriage.

Critical Thinking: Write Your Opinion

19) How can keeping a positive attitude help you solve a problem?

20) What could you do to help a friend who is suffering from abuse?

Test Taking Tip Before you begin a test, look it over quickly. Try to set aside enough time to complete each section.

Deciding for Yourself

Planning a Fitness Program

Do you have a fitness or exercise program? You can help keep your body healthy by exercising regularly. Anyone can do it. It takes some planning and making a decision to stick to your plan. Here are some steps for setting up an exercise program.

- Think about a physical activity you enjoy. It can be biking, swimming, or playing a sport. For example, if you like hockey, skating is important.
- Plan a time in your day that you could do this activity regularly. Maybe you would have time after school or in the early evening. Think about the amount of time you have and where you will do the activity.
- Set some goals for yourself. For example, you might want to walk or run a certain distance in fifteen or thirty minutes. Make your goal realistic and start slowly.
- Think about ways to reward yourself when you reach a goal. This is a good way to celebrate your progress.
- Make friends with other people who enjoy the same activity. Friends can help one another stick to their exercise program.

Questions

1) What is a physical activity that you enjoy doing?
2) What kind of a goal might you set for yourself with this activity?
3) How do you think doing this activity three times a week would help your physical health? How might it help your emotional health? Your social health?
4) What kind of warm-up or cool-down activity would you do before you exercised?

Unit Summary

■ Tiny cells make up all living things. Your body is made of trillions of cells.

■ Your body is organized into cells, tissues, and organs. Tissues that work together to do a similar job are called organs. Many organs work together in a body system.

■ Your body systems include the skin, skeletal, muscular, digestive, excretory, respiratory, circulatory, nervous, endocrine, and reproductive systems.

■ Body systems work together to provide your cells with oxygen and nutrients.

■ Caring for your skin, hair, nails, teeth, eyes, and ears is a part of hygiene.

■ Exercise can improve your physical, social, and emotional health.

■ An exercise program should include a warm-up period, an exercise period, and a cool-down period.

■ Most teenagers need eight or nine hours of sleep each night.

■ Each stage in the family life cycle brings new challenges to a family.

■ The family life cycle begins with marriage.

■ The second stage in the family life cycle occurs when a child is born into the family.

■ The third stage in the family life cycle is when children become teenagers and begin to prepare to live on their own.

■ The fourth stage in the family life cycle is when children leave home and begin their own lives.

■ The final stage in the family life cycle is learning to deal with aging.

■ Stress occurs in all families. In healthy families, members learn to work together to solve problems.

■ Family members often feel denial, anger, adjustment, and acceptance as they react to problems.

■ Violence and abuse are never acceptable in a family.

UNIT 1 REVIEW

Comprehension: Identifying Facts

On a separate sheet of paper, write the correct word or words from the Word Bank to complete each sentence.

WORD BANK		
plasma	extended	safety
cardiovascular	cells	acne
skeletal	epidermis	endocrine
bone	hygiene	abuse
physical	brain	

1) Your blood is made of cells and _____.

2) Hormones are chemicals released by _____ glands.

3) The cerebrum is the largest part of your _____.

4) When a muscle pulls a _____, it causes your body to move.

5) Your _____ system is made of 206 bones.

6) Your digestive system breaks food down into a form your _____ can use.

7) The _____ is made of dead skin cells.

8) _____ is blocked skin pores.

9) Caring for your skin, hair, nails, teeth, eyes, and ears is a part of _____.

10) _____ fitness is the condition of the heart, lungs, and blood vessels.

11) _____ fitness is your body's ability to exercise, work, and play without tiring easily.

12) A family that includes people from different generations is called an _____ family.

13) Every state has laws to protect children from _____.

14) Parents are responsible for the health, _____, and well-being of their children.

Comprehension: Understanding Main Ideas

Write the answers to these questions on a separate sheet of paper. Use complete sentences.

15) Describe the five stages in the family life cycle.

16) Explain how exercise affects physical, social, and emotional health.

17) What does the word *family* mean?

18) Choose two body systems and explain how they work together to do a job in the body.

Critical Thinking: Write Your Opinion

19) What are two things you can do to improve your personal hygiene?

20) What role does your family play in your everyday life?

> *A strong positive mental attitude will create more miracles than any wonder drug.*
> —Patricia Neal

Unit 2

Mental and Emotional Health

For most people, some days are better than others. You can have good days and bad days. Sometimes outside events influence the kind of day you have. But your thoughts, emotions, and beliefs also play a role. Learning how to maintain good mental and emotional health is important. It can help you enjoy the good days more and not be as upset by the bad days.

Your beliefs about yourself affect what you do. They can have a big impact—either positive or negative—on your physical health. These beliefs also influence how you handle relationships with other people. In this unit, you will learn about mental and emotional health. You will also learn ways to create and maintain healthy relationships.

- ▶ Chapter 4 Emotions
- ▶ Chapter 5 Maintaining Mental Health
- ▶ Chapter 6 Relationships

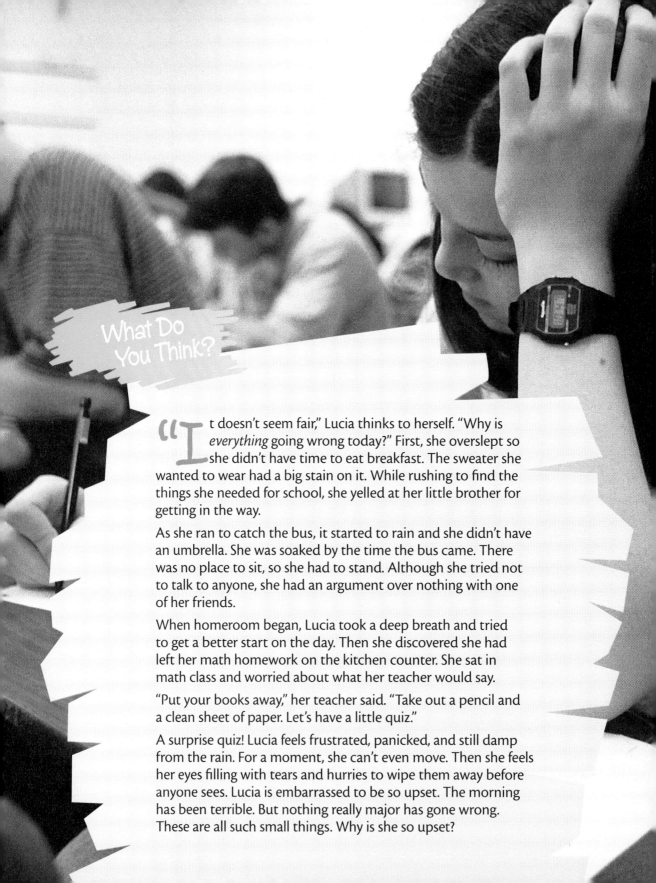

What Do You Think?

"It doesn't seem fair," Lucia thinks to herself. "Why is *everything* going wrong today?" First, she overslept so she didn't have time to eat breakfast. The sweater she wanted to wear had a big stain on it. While rushing to find the things she needed for school, she yelled at her little brother for getting in the way.

As she ran to catch the bus, it started to rain and she didn't have an umbrella. She was soaked by the time the bus came. There was no place to sit, so she had to stand. Although she tried not to talk to anyone, she had an argument over nothing with one of her friends.

When homeroom began, Lucia took a deep breath and tried to get a better start on the day. Then she discovered she had left her math homework on the kitchen counter. She sat in math class and worried about what her teacher would say.

"Put your books away," her teacher said. "Take out a pencil and a clean sheet of paper. Let's have a little quiz."

A surprise quiz! Lucia feels frustrated, panicked, and still damp from the rain. For a moment, she can't even move. Then she feels her eyes filling with tears and hurries to wipe them away before anyone sees. Lucia is embarrassed to be so upset. The morning has been terrible. But nothing really major has gone wrong. These are all such small things. Why is she so upset?

Chapter

4

Emotions

Being healthy involves both the body and the mind. Feeling happy can help your whole body feel well. If you are worried about something, you might not feel as well. Your mind affects the body. The body also affects the mind. For example, if you get enough rest, you are usually able to think clearly. If you are tired, you may not be able to think as clearly.

In this chapter, you will learn about keeping your mind and body healthy by understanding your emotions. You will find out what causes emotions. You also will learn how emotions affect your relationships with others and the way you behave.

Goals for Learning

▶ To learn what emotions are

▶ To describe some causes of emotions

▶ To explain how emotions affect relationships with others

▶ To explain how emotions are related to the way a person behaves

Lesson 1

Emotions and Their Causes

Emotions
Feelings

*E*very person has feelings. Your feelings make you different from every other person. Your feelings affect the way you react to things. Different people may react to the same thing in different ways.

What Are Emotions?

Your feelings are called **emotions**. Emotions are our reactions to events and experiences. They let us know when something is right as well as when something is wrong. Learning about your feelings can help you to understand yourself better.

An emotion is made up of two parts. One part is the physical reaction you have to a situation. For example, you might feel angry. Then your face might become flushed, or reddish. Your muscles may feel tight. You might even want to shout or hit something. The second part of an emotion is the mental side. The mental side of an emotion explains why you feel the way you do.

The Physical Side

Different kinds of physical changes can be part of emotions. For example, when you are excited, your blood may flow more quickly through your body. Then you might feel stronger and have more energy. If you are sad or afraid, the muscles in your legs might feel tight. You might feel like you are ready to run away. You might even cross your arms to protect your body. Physical changes get your body ready to act. Before you act, it is wise to think about why you feel the way you do.

Before you act, think about why you feel the way you do.

Stress
The way the body reacts to a change or to something that can hurt you

The Mental Side

Understanding why you feel a certain way can help you decide how to act. Suppose you are angry and act before you think about the reasons for your feelings. You might do something you will feel sorry about later. For example, if you say unkind things to a friend you might end your friendship. If you think before you act, you might behave differently.

What Causes Emotions?

Emotions start with thoughts. When we think about something that happens, we may feel a certain way about that event. Our thinking triggers an emotion. **Stress**, fear, and anger are common emotions.

Stress

When we feel threatened, we may experience stress. Stress is a state of physical and emotional pressure. Stress can be "good" or "bad." For example, you may be excited about a project and do a good job on it. Then good stress helps you. Usually we hear about bad stress because its effects are harmful. Changing schools or the death of someone you love are examples of bad stress. Bad stress can interfere with healthy living. The emotions of fear and change are linked to bad stress. The following chart lists events that commonly cause stress for teens.

Common Stressful Life Events for Teens

• Death of a parent	• An outstanding personal achievement
• A visible deformity	• Being accepted to college
• Parents' divorce or separation	• Being a senior in high school
• Being involved with alcohol or other drugs	• A change in acceptance by peers
• Death of a brother or sister	• A change in parents' financial status
• Beginning to date	

ANCIENT IDEAS

Around 400 B.C., the Spartans of ancient Greece believed in physically strong people. Young boys were trained in sports. They were made to exercise so they became strong soldiers. They learned only how to fight and farm. They did not have any other schooling.

At the same time, the Greek people of ancient Athens believed in strong minds and bodies. They wanted their people to learn about history, science, and the arts.

Most cultures have followed the Athenians. They believe that the mind and body work together. Developing both the mind and the body helps a person be healthier.

Fear

When you are afraid, you feel uncomfortable and nervous. You want to get away from whatever is frightening you. Fear is useful when it helps you protect yourself. For example, you may be playing ball and a teammate shouts, "Watch out!" You may feel fear and duck your head. Sometimes you might feel afraid without knowing why. If someone tries to get you to do something you think is wrong, you might feel afraid. Your fear may help you get away from the situation.

Anger

Anger might cause you to react too strongly. You might start a fight or an argument. If someone bothers you, you may want to fight back. When you feel anger, you can choose how to behave. You can speak or yell. You can complain to someone who can help you. You can hurt someone or step away until you calm down. Before you act, it is a good idea to think about what might happen later. Deciding how to act is not always easy.

Fitness Tip

Getting regular exercise can make it easier to deal with stressful situations.

Adapt
Change

The best way to express your anger is to say why you are angry. You could say, "I feel hurt when you tease me." Staying calm is a way to control anger. Self-talk helps, too. For example, it can calm your body to tell yourself, "Stay calm. Relax."

How Can You Change Your Behavior?

When you feel the emotions of stress, you can **adapt**, or change, the way you behave. Adapting means that you act in ways that solve problems. You adapt to cold weather by wearing warm clothing. You also can adapt to situations that cause stress. If you are worried about a test, you can study. If someone hurts your feelings, you can tell the person how you feel. You can ask the person to apologize. Keep in mind that the person might not agree to apologize. Nevertheless, by trying, you will feel that you have done all you can.

How can walking away from a fight help you feel better?

The emotions of relief and joy are a sign that the stressful situation is over. Relief can feel like a wave of good feelings. It ends the fear or anger that stress starts. Joy is a happy feeling. People show joy in different ways. Some people smile shyly. Others laugh out loud.

Feeling good is important for mental health. Normal, healthy living usually has some ups and some downs. Changes in emotions help you keep a healthy balance in life. It is important to recognize what your emotions are trying to tell you.

LESSON 1 REVIEW Write the answers to these questions on a separate sheet of paper. Use complete sentences.

1) What are two sides of an emotion?

2) What emotions prepare people to protect themselves?

3) Name one way you can deal with anger in a way that can help solve problems.

4) What emotions can be a sign that a stressful situation is over?

5) What is one way you could adapt your behavior if you are afraid about acting in the school play?

Lesson 2

Social Emotions

Social emotions
Emotions that have to do with relationships with others

Guilt
An emotion felt when a person does something wrong

People need other people. Many emotions begin with our relationships with family members, friends, and other people.

What Are Social Emotions?

Emotions that have to do with relationships with others are called **social emotions**. The most familiar social emotion is love. Love is a positive way of thinking, feeling, and acting toward another person. People feel different kinds of love. You feel love for parents, brothers, and sisters. You also feel love toward close friends. As you grow older, love can develop for a special person in your life.

Love begins with a baby's need to be near a parent or caregiver. As you grow and meet more people, you choose to have loving feelings toward others. Love involves sharing and respect between people. The person you love feels loving feelings for you, too.

What are some ways to show respect to a family member you love?

Love is important for healthy living. Painful social emotions develop when love is lost. You feel the emotion of **guilt** when you disappoint yourself and others. You may feel guilty when you do something you know is wrong. Guilt may not have signs that other people see. Guilty feelings may keep you from having good relationships with others.

You may feel love toward close friends.

Shame
An emotion that results from disapproval or rejection

Grief
A mixture of painful emotions that result from loss

Shame is an emotion you feel if someone does not approve of you. You might feel shame if someone rejects you. Suppose you make a mistake. Suppose you feel that people are making fun of you. You might feel embarrassed and ashamed. Shame sometimes has outward signs. People who are ashamed sometimes blush. When you blush, your face becomes red and warm. If you feel shame, you might perspire. You probably wish to get away from the situation.

Sometimes people lose close relationships. For example, a grandparent dies or a good friend moves away. A person you care about might stop caring for you. You feel **grief**. Grief is a mixture of painful emotions you feel when you have a loss.

Grief is a normal way to react to a loss. A grieving person needs to deal with his or her feelings. It is important to face the loss and talk about the pain it causes.

Action for Health

DEALING WITH LOSS AND DISAPPOINTMENT

How can you learn to deal with grief and disappointment? There are some things you can do to deal with grief in positive ways.

Accept the loss. You can keep photographs or objects to remind you of a person you lose. If you are disappointed about something, try writing about your feelings.

Set a time every day for grieving. Explain to others that you need some time alone. You may need to cry. Crying expresses sad feelings. It also helps bring relief. When time is up, put your grief away. Remind yourself that you can grieve again the next day.

Make a plan for living with your loss. Talking with others may be helpful. Set new goals. A goal might be to exercise every day. Think of new dreams for your life. You might even set a goal for something you want when you are older.

Reward yourself. When you reach a goal, be proud of yourself. You might want to tell someone else what you did. As you reach some goals, you can set other goals. Think about all the things you can look forward to in your life.

Depression
Extreme sadness

Grief includes **depression**. Depression is extreme sadness. People who feel depressed have low energy. They may feel angry and stay away from others. A grieving person may feel depressed for weeks or months or even longer.

People also feel grief when they have disappointments. For example, you may feel depressed and angry when you do not make a team. Not being chosen for a club can lead to depressed feelings. You may feel depressed when you do not meet one of your goals. If you feel depressed, it is important to tell your feelings to someone who cares about you. If friends feel depressed, it is important to listen to their feelings. Encourage them to express their feelings.

The best "cure" for depressed feelings is positive action and thinking. Sometimes, the depressed person needs the help of a mental health professional, such as a psychologist. Often, family members and friends can help the depressed person get over the pain of losses. Then the person can begin again to be active and to have good relationships.

Nutrition Tip

Take time to eat healthy meals when you are feeling sad or disappointed about something.

LESSON 2 REVIEW Write the answers to these questions on a separate sheet of paper. Use complete sentences.

1) When does love begin?
2) What emotion might you feel when you do something you know is wrong?
3) What signs might a person show if he or she feels ashamed?
4) What are some things that can cause grief?
5) What might you suggest to a friend who seems depressed?

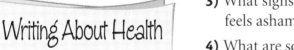

Writing About Health

Think about a goal you could set that could help you deal with a disappointment. List things you could do to reach your goal.

Lesson 3

Emotions and Behavior

Need
Something important or necessary to have

Emotions always lead to some way of behaving or acting. Sometimes controlling your feelings probably seems easy. Other times, you may feel "driven" by your feelings. Then you are more likely to act without thinking carefully. Acting without thinking first can have unpleasant results.

What Do Emotions Mean?

Emotions are feelings that seem to come from inside you. If you watch a sad movie, you may feel like crying. If you do well on a test, you may feel happy. When you watch a game, you may feel excited. You may feel like jumping up and shouting to cheer your team. Emotions are automatic responses to your thoughts and experiences.

Some emotions are signs of **needs**. You have a need when you are missing something important. If you do not have enough food, you get hungry. You have a need to eat. If you do not feel close to other people, you feel loneliness. You need friendship. When emotions show that you need something, you can do something about the problem. Hungry people think about food and how to get it. They may think of hardly anything else. Lonely people think about being close to others. They may feel sorry for themselves.

When an emotion signals a need, the healthiest response is to think about what you can do. You can do three things:

1. Identify what causes the problem.
2. Decide on a way to solve the problem.
3. Carry out your plan.

When you identify your needs, you can determine how best to meet them. Sometimes all you need to do is ask. If you need help with your homework, simply ask a friend.

You may feel happy and excited after a sporting event.

Emotions   Chapter 4   **93**

Being direct works best. Dropping hints and hoping others will guess what you need usually doesn't work. You may need to work harder to get what you need. If you need high grades to get a job, you may have to study harder.

Emotions can be a sign that you are feeling stress. When something you do not expect happens, you may feel upset. You might want to act but you do not know what to do. If a teacher announces a pop quiz, you may feel angry. You might feel afraid even though you have been studying. Having a surprise test can seem unfair. In such a situation, you may not have many choices. You probably will take the test instead of running from the classroom. Maybe stress will help you work harder and do better on the test.

Emotional Action

The outcome of an emotional experience depends on what you do. Recognizing how you feel is important. Everyone feels angry or afraid at times. You can control the way you act when you have such feelings. Then you will be better able to deal with problems. Behaving in a reasonable way will help you feel good about yourself.

Healthy Subjects
Literature

Toni Morrison's novel *The Bluest Eye* tells the stories of two African American girls. The girls come from different families. One girl, Pecola Breedlove, has parents who are angry. Her parents abuse her. Other children laugh at her. She feels ugly for being black. Pecola wishes she were white. She believes that having white skin would make her happy. She even prays for blue eyes. Pecola's sadness just makes her ill. As time passes, her health gets even worse.

The other girl, Claudia McTeer, feels very loved by her family. She likes the way she looks. She accepts herself. She has dreams about her future. Claudia works hard so her dreams may come true someday. Accepting herself helps her stay healthy.

MENTAL HEALTH ASSISTANT

Do you think of yourself as caring? Would you describe yourself as trustworthy? Then you might want to become a mental health assistant. Mental health assistants work closely with mental health counselors. They help other people solve problems. Counselors help people handle many different kinds of situations. For example, some people have problems with their marriages. Others have trouble getting along with family members or friends. Sometimes people have problems that are related to school or to their jobs. People may need help handling stress. Some people do not think well of themselves. Others may be so depressed that they want to commit suicide. Mental health assistants need to be caring. They need to be able to gain people's trust. Good communication skills are important. Most employers prefer mental health assistants who have had college courses that teach "people skills" or psychology.

Compromise
An agreement in which both sides give in a little

Suppose that you want to go out with your friends. Your parents are not sure they want you to go. You feel angry at your parents. You are afraid that your friends will think you are a baby if you cannot join them. You can choose how to behave. You can have an angry outburst at your parents. Such behavior might hurt your parents' feelings. It is not likely to bring the results you want. If you think before you act, you might take more positive steps. For example, you could set a time to talk with your parents. You could explain where you want to go. You could tell them who will be with you. If you speak calmly, your parents will be more likely to listen to you. You also could listen to your parents' wishes and concerns. Acting in a reasonable way may lead to a **compromise**. A compromise is an agreement in which both sides give in a little.

Emotions Chapter 4

Keep in mind that all emotions are normal. Emotions are not "good" or "bad." How you handle an emotion is the important thing to your mental health. First you need to pay attention to the emotions you are feeling. Think about why you feel the way you do. Then you will be ready to act in a positive way. You might be able to change a situation that is causing you to feel afraid or angry. For example, you may be afraid of taking a history test. You can talk to your teacher. He or she can give you some ideas for studying in a better way.

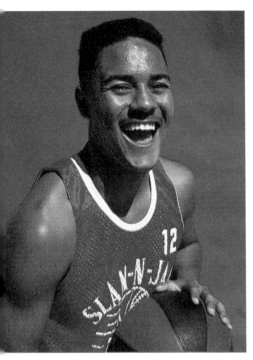

Playing sports, such as basketball, can help you relax.

Expressing your emotions can help you deal with them. You can express your emotions in different ways. You can talk with others about your feelings. You can write about your feelings. You can ask other people to help you meet your needs. For example, you can ask parents or teachers to help you make a decision.

What physical action might you choose to do to deal with stress in a positive way?

Sometimes taking physical action can help you deal with emotions. You might pat yourself on the back. You might jump in the air or sing a song. Doing physical activities, such as running or dancing, can help relax you. Then you can deal with your emotions in a calm way. Sometimes people feel that they cannot handle their emotions themselves. In such cases, talking with a mental health professional is the best thing to do.

LESSON 3 REVIEW Write the answers to these questions on a separate sheet of paper. Use complete sentences.

1) What are emotions?
2) What is an example of positive behavior that can come from stress?
3) How can thinking before you act be important?
4) What are two positive ways you can deal with an emotion?
5) What is an example of a situation in which a compromise might be helpful?

Chapter Summary

- Emotions are automatic responses to thoughts and experiences.

- Acting without thinking first can have unpleasant results. Thinking first can help a person act in a reasonable way.

- Some emotions are signs of needs people have, such as food or friends.

- When emotions signal a need, a person must decide how to solve the problem.

- Stress can help a person work harder to solve a problem.

- Recognizing feelings can help a person control his or her actions.

- Behaving in a reasonable way helps a person feel good about himself or herself.

- Angry outbursts are not likely to bring positive results. Acting in a reasonable way may lead to a compromise.

- How a person handles emotions is important to mental health.

- Understanding the reasons for emotions can lead to positive behavior.

- People can often change situations that are causing them to feel angry or afraid.

- Expressing emotions can help a person deal with them. Talking with a mental health professional is important when people are unable to handle emotions themselves.

- Physical activity can help a person deal with emotions in a positive way.

Chapter 4 Review

Comprehension: Identifying Facts

On a separate sheet of paper, write the correct word or words from the Word Bank to complete each sentence.

WORD BANK	
adapt	guilt
compromise	need
depressed	shame
emotions	social emotions
grief	stress

1) _____ are a person's reactions to events and experiences.

2) You can _____ the way you behave to solve problems.

3) You may feel the emotion of _____ when you do something you know is wrong.

4) Emotions that have to do with relationships with others are called _____.

5) A _____ is an agreement in which both sides give in a little.

6) A person who feels _____ feels very sad and usually has low energy.

7) When emotions show that you have a _____, it is important to do something about the problem.

8) _____ is a mixture of painful emotions a person feels because of a loss.

9) A person might feel the emotion of _____ if he or she is rejected by someone else.

10) _____ is a state of physical and emotional pressure.

Comprehension: Understanding Main Ideas

Write the answers to these questions on a separate sheet of paper. Use complete sentences.

11) What kinds of physical changes can be part of emotions?
12) Why might some people be frightened of some of their emotions?
13) What are some ways you can choose to behave when you are angry?
14) What are three reasons people behave the way they do?
15) How can the emotion of fear be useful?
16) What are two ways you can express your emotions?
17) What are some ways to deal with feelings of depression?
18) What are some signs of love?

Critical Thinking: Write Your Opinion

19) Why do you think dealing with stress and other emotions is important to your physical and mental health?
20) What could you do to help a friend who is feeling the emotion of grief?

Test Taking Tip When you take a test, be sure to read each question carefully before answering.

Chapter 5

Maintaining Mental Health

Many things affect your mental health. Getting to know yourself helps your emotional health. Becoming aware of your values and beliefs can also help your well-being. You can do many things to help keep yourself mentally and emotionally healthy.

In this chapter, you will learn about healthy ways to deal with frustration and stress. You will find out ways to handle peer pressure. You also will learn about eating disorders and how they affect health.

Goals for Learning
▶ To learn ways to handle frustration
▶ To identify healthy ways to cope with stress
▶ To identify ways to deal with peer pressure
▶ To describe three kinds of eating disorders

Lesson 1

Managing Frustration

Aggression
Any act that is meant to harm someone

Frustration
An unpleasant feeling that happens when goals are blocked

Everyone has needs. Each person also has goals he or she wants to meet. When you are hungry, you need to eat. When you are lonely, you want to be with people. When you want to do well in school, you study hard.

How Can You Deal With Frustration?

Sometimes you feel the unpleasant feeling called **frustration**. Frustration happens when you are blocked from meeting your goals. Sometimes someone or something else gets in the way of your goal. For example, Hank wants to take a shower. His brother Ross gets in the shower first. Ross's action makes Hank feel frustrated.

You feel frustrated when something you do gets in the way of your goal. Part of Hank's problem was that he overslept. If he had awakened on time, he could have used the shower first. You frustrate yourself when you do not plan ahead. You frustrate yourself when you make mistakes in meeting your goals. You also feel frustrated when you do not have everything you need to meet a goal.

Responses That Are Not Helpful

People react to frustration in different ways. Often people react in ways that are harmful. Such responses do not get rid of things that block goals.

Aggression is the most common response to frustration. Aggression is any act that is meant to harm someone. For example, Hank's little sister gets in his way. Hank shows aggression when he yells at her. The way Hank acts does not help the problem. Aggressive actions don't solve the problem.

Another reaction to frustration is acting childish. People may whine when they don't get their way. Sometimes people pout or stamp their feet and cry. They may bully other people. Such actions do not help you get what you want.

Cope
Deal with a problem

Withdraw
Pull away

If you feel frustrated a lot, you may **withdraw**, or pull away. People sometimes withdraw if they cannot deal with problems. They may withdraw if they feel frustrated. People who withdraw sometimes give up at the first sign of a problem. They stop trying to deal with the problem. Marty has trouble with math. He has to work harder to learn math than his classmates do. When he is frustrated by learning math, he stops work early. He complains that he'll never figure it out. Sometimes others try to help or explain the math. Marty acts impatient and does not listen. Marty still feels frustrated. Withdrawing does not help a person deal with a problem.

Responses That Are Helpful

The positive way to respond to frustration is to **cope**. When you cope, you handle the problem. You get rid of whatever is causing your frustration. Coping is not always easy. It does make a positive difference. Coping helps you deal with a problem.

One way to cope with frustration is to take another look at your goal. Think about what is blocking you from meeting the goal. Maybe the goal is too far out of reach. There may be no way to meet it. For example, Bill likes playing softball, but he is not a fast runner. His goal of making the school softball team is blocked. He cannot run fast enough to make the team. Bill sets a new goal. He decides to play softball with friends on weekends. Bill is able to meet his new goal. Making sure your goals are possible is a good idea.

People react to frustration in different ways.

Another way to cope with frustration is to work harder to reach your goals. For example, Maria wants to go to computer camp this summer. Right now, her grades are not high enough to get into the camp. She sets a goal to raise her grades. Maria decides to work harder in school. She talks with her teachers about what to do. She gives up some other activities and spends more time studying. She probably will be able to get better grades. Then she will be able to go to computer camp this summer.

In some cases, there is another way to cope with frustration. You figure out what is blocking you from reaching a goal. Then you can figure out a way to solve that problem. For example, Nina wants to buy a new camera. She has not saved enough money to buy the one she wants. She talks with the owner of the camera store about her problem. Nina and the owner agree that she can trade in her old camera. In this way, Nina will get a lower price on the new camera. Now Nina can afford to buy the camera she wants.

What reasonable goal can you set for yourself this week?

In most cases, people need to act in several ways when they feel frustrated. Combining the ideas given above will help you cope with frustration. These ideas work because they solve the problems that are preventing you from reaching your goals.

LESSON 1 REVIEW Write the answers to these questions on a separate sheet of paper. Use complete sentences.

1) What causes frustration?

2) How does acting in an aggressive way increase frustration?

3) What are two ways of acting that show aggression?

4) Think about a person who gives up when there is a problem. What word describes the way the person is acting?

5) Suppose you are frustrated because you did not make the school band. What are some things you can do to cope with the frustration you feel?

Lesson 2

Managing Stress and Anxiety

Hassle
A small, annoying event or problem

Self-esteem
The way a person feels about himself or herself

Threat
A situation that puts a person's well-being in danger

You may recall reading about stress in Chapter 4. Stress is a state of physical and emotional pressure. Some physical activities cause stress. Running to catch the bus or straining to lift a heavy object are stressful. Experiences that take mental effort also cause stress.

How Can You Learn to Manage Stress?

Stress in itself is not bad. In fact, it helps you do certain things. For example, stress helps you work quickly to finish a paper for school. It gives you energy to run fast to catch the bus. Stress keeps you alert. It helps you take the right action. Too much stress can be harmful to your health.

Causes of Stress

Threats are a common cause of stress. Threats are situations that seem to be dangerous to you. People feel threatened when they cannot meet their basic needs. For example, people are threatened when they do not have enough food. Not having shelter is a threat. People also feel threatened when they do not feel safe. Sometimes stress is caused by threats to **self-esteem**. Self-esteem is how a person feels about himself or herself. Your self-esteem is threatened when someone puts you down or criticizes you.

What is a positive way you have dealt with a hassle?

Other common causes of stress are major life changes. These changes are losses, illness, and injuries. People need to learn new ways of acting when they face major changes. People also need to deal with **hassles**. Hassles are small, annoying events or problems. Think about hassles that make life hard. You might oversleep or miss your bus. You might spill something or drop your books. Someone might get angry at you. One or two hassles are not stressful enough to harm your health. When a number of hassles happen at the same time, they cause stress.

Stress response
The body's reaction to stress

Physical Signs of Stress

When you face stress, you may notice changes in your body. For example, suppose your teacher wants to talk with you. You are afraid you may hear bad news. You feel stress. Most of your response may be automatic. Your body's response to stress is called the **stress response**. There are three parts of the stress response. During the first part, your heart beats faster. Your breathing speeds up. You perspire and your muscles get tight. You may blush or cry out. Sometimes people have headaches or stomach pains. They may feel nervous. Some people may not be able to sleep well.

The second part of the stress response takes one of two forms: fight or flight. When something threatens you, you may want to fight to defend yourself. Fighting may be using words to stand up for your beliefs. The other response is flight. When you have this response, you want to flee, or run away. You leave the situation. The fight-or-flight response is natural. It happens without thinking.

Sometimes stressful situations happen over and over. Then the third part of the stress response happens. You feel exhausted, or very tired. You use a lot of energy dealing with stress. Dealing with stress over a long period of time may lead to illness. People may develop health problems such as ulcers or high blood pressure.

Health Tip

Taking deep breaths and breathing in and out slowly will help relax you when you feel stress.

Healthy Subjects: Geography

SEASONAL AFFECTIVE DISORDER

Scientists have found that Seasonal Affective Disorder (SAD) affects five million Americans. The condition is very stressful. People feel tired and depressed. They may be unable to sleep. They usually feel anxious. Less daylight during winter months causes SAD. Northern parts of the United States have more cases of SAD. Alaska has the greatest number of cases. Florida has the fewest. Less daylight seems to reduce the amount of an important chemical in some people's brains. Treatment with light can increase the amount of the chemical. People sit in front of a light box for thirty minutes a day. When they do, they have fewer signs of SAD.

Anxiety
An unpleasant feeling like fear but without reasons that are clear

Managing Stress

After you first respond to something stressful, your body calms down. Then you can decide what to do. For example, suppose you walk into a dark room and hear a noise. You may jump back. Then you decide to turn on a light to see what is making the noise. You find the noise was nothing to worry about.

Another way to manage stress is to prepare yourself for it. Eating right, exercising, and getting enough sleep help keep your body healthy. You can do things to strengthen your feelings and your thoughts. For example, do something every day that makes you feel good about yourself. You might want to do something you enjoy or finish something you started. Developing your talents is important. Learning how to handle your own problems will help you feel strong. You might ask others to show you how to do things yourself. Don't be afraid to ask others for help when you need it.

How Can You Deal With Anxiety?

You have learned about fear, anger, joy, and grief. However, some feelings are hard to describe. You may have mixed emotions about some things. You may not even be sure why you feel good or bad. One unpleasant feeling that is hard to describe is **anxiety**. Anxiety is a feeling something like fear. Yet the reasons for anxiety may not be clear. People who are anxious do not really feel afraid. They feel *like* they are afraid. They often feel nervous or have trouble sitting still. People who are anxious feel uncomfortable. Anxiety is an upsetting feeling.

Anxiety is worse when you don't know exactly what is causing it. Instead of finding the reason for anxiety, you may just try to get rid of it. For example, Keith feels anxious about getting ready for school one morning. He probably feels anxious because he did not study enough for his French test. He doesn't try to find the reason for

Do something you enjoy to reduce stress.

Writing About Health

List three situations that often cause you to feel stress. Next to each item, write about a way to deal with the stress in a positive way.

his anxiety. Instead he tries to get rid of the anxiety by staying home from school. Later when he feels better he must still go to school. He does not understand why he feels anxious. When he gets to school, he still is not ready to take the test.

The best way to deal with anxiety is to find the reason for it. Then you can face the problem and deal with it. When you deal with the problem that is causing anxiety, you will feel relief.

LESSON 2 REVIEW Write the answers to these questions on a separate sheet of paper. Use complete sentences.

1) What are some positive effects of stress?

2) What are two common causes of stress?

3) How does the stress response affect your body?

4) What are three signs of anxiety?

5) What are three ways to help you manage stress?

technology

BIOFEEDBACK

When you hear *biofeedback*, do you think of science fiction? Biofeedback is not science fiction. It is used to help treat illness or anxiety. Technicians place special sensors on a person's head or muscles. Computers measure muscle activity. Changes in muscle activity show on a computer monitor. The monitor shows how much stress a person is feeling. People learn to control pain by using positive thoughts. Biofeedback is used to treat long-term pain. It also is used to treat many other disorders such as headaches, epilepsy, and sleeplessness. Athletes use biofeedback to control anxiety before games. Biofeedback is not always a sure cure. For some people, though, it seems to work.

Maintaining Mental Health   Chapter 5

Lesson 3

Responding to Peer Pressure

Peer
A person in the same age group

Peer pressure
The influence people of the same age have on one another

When you were younger, you spent most of your time with your family. Your family is still very important to you. As you grow older, your friends are also becoming an important part of your life.

Friends keep you company. They laugh at your jokes. Friends listen to you when you need to talk. They help you do things. Groups of friends give one another a sort of home. The people in the group have a feeling of belonging. Some goals are easier to reach when you are part of a group. Your friends help you do things you could not do alone. For these reasons, you probably belong to some groups. The groups may be teams or clubs. You may have a special group of friends. Your friends and other people who are close to your age are called your **peers**.

How Does Peer Pressure Work?

Belonging to a group gives you a good feeling. In return for that feeling, others in the group expect something of you. They expect you to pay attention to them. They expect you to support things they do and say. Peers often influence things you say and do. For example, you may wear the same styles of clothing as your peers. You may join the same activities. The influence your peers have on you is called **peer pressure**. Both words and actions are part of peer pressure.

Many times peer pressure is reasonable. It can even be helpful. Your friends and teammates help you meet your goals. They often encourage you to do good things. Lee just moved to a new school. Her new friends make visits to older people at a nursing home. They asked Lee to join them. Lee decided to go with her new friends.

What is one way you have resisted peer pressure?

Lee found that she enjoyed talking with older people. The visits made her feel good about herself. Peer pressure helped Lee to do something worthwhile. She did something that went along with her own beliefs. In this case, peer pressure was positive.

Resist
Act against

Sometimes you may get harmful peer pressure. Then you feel pressure to act in ways that are wrong for you. For example, Lucy planned to study in the library after school. Her friends asked her to go to a movie instead. She was afraid her friends would be angry if she studied. Even though schoolwork was important, she went to the movie. She gave in to the peer pressure by doing something that was against her beliefs.

David had history homework. The night before the homework was due, David's friend Tom called. Tom said that he had not had time to do the homework himself. He asked if he could copy David's answers. David was afraid of losing Tom as a friend. Yet David knew that cheating was wrong. He told Tom that he did not feel right about cheating. He did not let Tom copy the answers to the homework.

Peers can be very hard on you if you do not go along.

Peer pressure can influence a person to do something that is wrong. For example, someone might ask you to speed down a hill on your bike. Someone might dare you to do something that might hurt you. Remember that you need to act in ways that keep you safe. You need to be strong enough to **resist** this kind of peer pressure. When you resist something, you act against it.

People often use peer pressure to get you to go along with the decisions of the group. Groups can be very hard on you if you do not go along. If you do not agree with the others, they may push you out of the group. At such times, you need to stand up for your own beliefs.

Maintaining Mental Health Chapter 5

How Can You Resist Peer Pressure?

Learning how to deal with peer pressure is an important part of growing up. Choosing whether to go along with others is not always easy. You might be afraid you will lose friends if you do not follow them. You might be afraid that you will no longer be part of a group.

Keep in mind that true friends will respect your feelings. Your ideas will be important to them. They will understand that you need to decide for yourself. True friends will still be friends if you disagree with them.

If people pressure you to do something wrong or dangerous, you *must* resist. You can handle this kind of peer pressure in different ways. Here are some things you can do to resist peer pressure.

- Ask questions. You have the right to know why people want you to go along with them. You may not agree with their reasons. Remember that you have the right to make up your own mind.

- Express your feelings. Starting explanations with "I feel . . ." or "I think . . ." is helpful. Words such as "You just want to get me in trouble" might cause a fight. Instead, talk about your own concerns. You might say "I think I would get into trouble. I'm not going along with you."

- Try to find someone else in the group who agrees with you. It is helpful if at least one other person sees it your way. Try to guess who would be the most understanding. Then ask that person to support you. You may not be able to figure out who would be on your side. You still can give others the idea that you are not alone. You might say "I'm not the only one who thinks this is a bad idea. I might just be the only one who will say so right now."

- You do not always need to explain your reasons. Say "no" or "I don't want to." Speak in a friendly way. Keep your voice calm. Then just walk away. People who are really your friends will not argue with you.

Action for Health

PRACTICING RESPONSES TO PEER PRESSURE

The best way to respond to peer pressure is to stand up for your rights. At the same time, respect the rights of others to choose differently. It is important for your body language to match your words. That is, you have to look like you believe what you are saying. Act sure of yourself and your decisions. Here are some things you can do when you respond to peer pressure.

- Stand straight and tall. Stand two or three feet away from the other person.
- Look into the other person's eyes. Keep your eyes steady.
- Have a pleasant expression on your face. Your expression can still make it clear that you mean what you say.
- Stay calm. Try not to look nervous. Keep your hands relaxed. Do not slump your shoulders or stuff your hands in your pockets. Try not to show fear.
- Practice your responses by looking in the mirror. If you practice your responses ahead of time, you will handle peer pressure more smoothly.
- If nothing else works, leave the group. You may only need to leave for a short while. Later, people may realize that you were right. Others may decide not to pressure you any more. You may need to leave the group for good. Then remember that you did so for the right reasons. You do not need friends who do not support you. It is better to have friends who think about what is best for you.

LESSON 3 REVIEW Write the answers to these questions on a separate sheet of paper. Use complete sentences.

1) What are three ways peers are important to you?
2) What are two examples of positive peer pressure?
3) What are two examples of harmful peer pressure?
4) What are three ways you can resist peer pressure?
5) What can you say to a friend who keeps pressuring you to do something wrong?

Maintaining Mental Health Chapter 5

Lesson 4

Eating Disorders

Anorexia
An eating disorder in which a person chooses not to eat

Body image
The way each person sees himself or herself

Eating disorder
A health problem in which a person loses control over eating patterns

During the teenage years, people develop their **body image**. Body image is the way each person sees himself or herself. Sometimes people see themselves in a positive way. In many cases, people have a negative body image.

People often get ideas about body image from the media. Most models are thin. Many young people think of a very thin body as a "perfect body." Yet few people really have this body type. When people compare themselves to an image of a "perfect body," they feel bad about themselves.

Teenagers must understand that their own bodies might never be like those of models. People naturally have bodies of many different types. A healthy, happy person looks better than a person who is unhealthy and unhappy.

What Are Eating Disorders?

Different people have different eating habits. Most people are able to control the way they eat. Some people have serious **eating disorders**. Eating disorders cause people to lose control over their patterns of eating.

Anorexia

A common eating disorder in the United States is **anorexia**. Anorexia is a serious disorder that happens when a person chooses not to eat. Anorexia results from emotional problems. The disorder affects girls more often than boys. Usually, people who have anorexia think they are too fat. They might begin by dieting. They eat less and less. Soon they feel full after only a few bites of food. They lose too much weight. Often, people with anorexia exercise a lot to lose more weight.

What can you say to a friend who is thin but always complains about being fat?

Anorexia can cause heart problems. It can even cause death. People with anorexia need to be treated by a doctor. They may need to spend time in a hospital for treatment. They also need counseling to deal with the emotional causes of the disorder.

Bulimia
An eating disorder in which a person eats large amounts of food and then vomits

Bulimia

Another common eating disorder is **bulimia**. A person who has bulimia eats large amounts of food within a short time. Then the person vomits or takes laxatives. In this way, the person gets rid of food before it is digested. Bulimia affects girls more often than boys.

Bulimia causes many health problems. The vomiting causes tooth decay because stomach acids eat away at the teeth. Many people with this disorder keep their normal weight. They might appear to be healthy. Others lose too much weight. People with this disorder do not get enough nutrients to stay healthy. They may develop heart conditions. Bulimia can lead to death. Bulimia is caused by emotional problems. People with this disorder need psychological treatment. They also need medical treatment to deal with their physical health problems.

People with anorexia think they are too fat.

Then and Now

TREATMENT OF MENTAL ILLNESS

People have always been concerned about mental and emotional disorders. Ancient peoples thought evil spirits caused the problem. Holes were drilled in the person's head to let the spirits escape. Years later, healers performed magic to drive out devils. During the Middle Ages, people with mental illness were often tortured or starved to get rid of demons. At other times, people were locked up in insane asylums. These were buildings used to keep people with mental illness away from others.

Today, mental and emotional disorders are better understood. Counseling and modern medicine can be used to treat these disorders.

Maintaining Mental Health Chapter 5

Careers

RESIDENT ASSISTANT

Resident assistants work with patients in hospitals, mental health settings, and nursing homes. When patients need help, assistants answer the call. They may check patients' temperatures. They may measure blood pressure, pulse, and breathing rate. Sometimes they bathe patients. Resident assistants help people who must stay in bed or a wheelchair. They serve meals and help some patients eat. Resident assistants listen to patients and help lift their spirits. Sometimes resident assistants are called nurses' aides or nurse assistants. They may get on-the-job training or go through a training program.

Overeating

For some people, overeating becomes a regular practice. They eat a large amount of food. They eat a lot of snacks. They often eat when they are not hungry. They seem to lose control over the way they eat.

Overeating can cause a person to be overweight. It can also lead to heart disease and many other health problems. People who overeat do not always eat healthy foods. They may not be getting good nutrition.

A person who overeats regularly should see a doctor. A doctor can help the person learn how to eat properly and lose weight. Often overeating is connected to emotional problems. Counseling can help a person deal with the causes of overeating.

Nutrition Tip

Eating plenty of fruits and vegetables can help you control your weight in a healthy way.

LESSON 4 REVIEW Write the answers to these questions on a separate sheet of paper. Use complete sentences.

1) Where do many people get ideas about their body image?

2) How does anorexia affect health?

3) What pattern of eating is part of bulimia?

4) What two kinds of treatment can help eating disorders?

5) How can you develop a good body image?

Chapter Summary

- People become frustrated when they are blocked from achieving a goal.

- Frustration often causes people to respond with aggression. Aggressive actions do not solve problems.

- People who don't want to deal with their problems may withdraw.

- Coping is dealing with a problem. One way to cope with frustration is to set reasonable goals.

- A person can figure out what is blocking him or her from reaching a goal. Then the person can find ways to solve the problem.

- Working harder to meet goals can help a person cope with frustration.

- Stress can be good or bad. It can give you energy to do things. Too much stress can be harmful.

- Threats, life changes, and hassles cause stress.

- Physical responses, such as a fast heartbeat, are part of the stress response.

- Wanting to fight or to flee is the second part of the stress response.

- Keeping healthy can help a person manage stress.

- Acting in ways to strengthen thoughts and feelings can help a person manage stress.

- The best way to deal with anxiety is to find the reason for it and deal with it.

- Peers often influence the way people act. Peer pressure can be positive or harmful.

- Asking questions, expressing your feelings, and saying "no" are some ways to resist peer pressure.

- People who have eating disorders cannot control the way they eat. Anorexia and bulimia are common eating disorders.

- People with eating disorders need treatment for their physical and emotional health.

- People who regularly overeat may become overweight and damage their health. Overeating is often connected to emotional health. A doctor can help people learn how to eat healthy foods and lose weight.

Chapter 5 Review

Comprehension: Identifying Facts

On a separate sheet of paper, write the correct word or words from the Word Bank to complete each sentence.

WORD BANK	
aggression	peer
anorexia	peer pressure
anxiety	resist
body image	self-esteem
bulimia	stress response
cope	threat
eating disorder	withdraw
frustration	

1) A health problem in which a person loses control over his or her eating is called an _____.

2) _____ is an unpleasant feeling that happens when a person is blocked from meeting goals.

3) Sometimes people pull away, or _____, when they feel frustrated.

4) The _____ is the body's physical signs of stress.

5) _____ is how a person feels about himself or herself.

6) People who eat large amounts of food and then vomit have an eating disorder called _____.

7) _____ is any act that is a common response to frustration.

8) _____ is a feeling like fear for which the reasons are not clear.

9) _____ is the way a person sees himself or herself.

10) A _____ is a person in the same age group.

11) When you _____ something, you act against it or withstand it.

12) A _____ is a situation that seems dangerous.

13) _____ is the influence people of the same age have on one another.

14) When you _____ with a problem, you deal with it.

15) A person with _____ chooses not to eat.

Comprehension: Understanding Main Ideas

Write the answers to these questions on a separate sheet of paper. Use complete sentences.

16) What physical reactions are part of the stress response?

17) What are three harmful responses to frustration?

18) What are two ways in which peer pressure can be helpful?

Critical Thinking: Write Your Opinion

19) How can learning to set reasonable goals help you?

20) How can resisting peer pressure affect your self-esteem?

Test Taking Tip Sometimes it is easier to learn new vocabulary words if you make them a part of your speaking and writing in other discussions and subject areas.

Chapter

6

Relationships

Emotionally healthy people have good relationships with themselves and others. Getting to know yourself and thinking about your values can make a difference in your well-being. Having healthy relationships can help your outlook on life. It can have a positive effect on your mental health.

In this chapter, you will learn about being a friend to yourself. You will find out about things you can do to make friends with others. You will also learn about relationships and what makes them healthy.

Goals for Learning

▶ To learn how to be a good friend to yourself

▶ To explain what you can do to make and keep friends

▶ To describe what makes a relationship healthy

Lesson 1

Being a Friend to Yourself

Bond
An emotional feeling of closeness

Think about your friends. You probably are different from one another in many ways. Yet you probably have many things in common. You may be interested in the same things. You may laugh at the same things. You and your friends have a **bond**. A bond is an emotional feeling of closeness.

Your friends help you meet some needs. One of these needs is the need to belong. Friends help you look at things in new ways. They listen to you and give you support. Sometimes they help you decide what do. Friends share your happy feelings and your sad ones. They also allow you to do good for others.

Why Should You Be Your Own Best Friend?

Before you can be friends with others, you must be a friend to yourself. There are two things you can do to be a friend to yourself. First, you can talk nicely to yourself. Second, you can treat yourself kindly.

Talking Nicely to Yourself

Think about your self-talk. Your self-talk is what you say to yourself. Are you friendly to yourself? Do you tend to put yourself down?

You might use the words "should," "ought," or "must" too often. You might say "I should be the best student in the class." You may not be able to reach such a goal. Yet you do not need to put yourself down. You can change your self-talk in a positive way. You could say "I study hard in school. I work as hard as I can. I am a good student."

What are some things you can say to yourself after you do well on a test?

You might feel you don't do something as well as someone else. Instead, feel proud of what you do. Try not say things such as "I will never be as popular as he is." Instead, you could say "Acting polite and kind to others helps me make friends. My friends and I are important to one another."

Relationships   Chapter 6   **119**

Health Tip

Take time for yourself. Then you can get to know yourself better. You can think of all the things you like about yourself.

You might blame yourself for something that is not your fault. Some teenagers say "If I were better, my parents would not be getting a divorce." Instead of putting themselves down, the teenagers could say "I feel sad about my parents' divorce. I know it is not my fault."

You might make a big deal out of something that goes wrong. You might say "I really blew that test. Nothing is going right." You can change your way of thinking. You could say "I did not do well on that test. I can study harder. I can do better on the next test."

You might sometimes think too much about things you don't like about yourself. You might say "I don't know enough about interesting things. I'm boring." Thinking about things you do like about yourself is more useful. You could say "I'd like to learn more about that subject. I'll read about it."

Using words such as "never" and "always" can lead to harmful self-talk. You might say "I'm always messing up." It would be more useful to say "I made a mistake. Making mistakes is okay. I'll try to learn from my mistake."

To make your self-talk helpful, think about the kinds of mistakes you make. Try to catch yourself if your self-talk is a put-down. Then say something in a positive way instead. Remember to compliment yourself whenever you feel proud of what you do.

Remember to compliment yourself when you look or feel good.

Action for Health

CREATE YOUR OWN SUPPORT SYSTEM

An important part of mental health is having support. You need to be able to provide support for yourself. You can get ready ahead of time for times when you feel down.

Make a list of things you can do to deal with problems. For example, write down ideas such as "talk it over," or "ask for help." Make sure the names of your close friends and their phone numbers are on your list. Look at your list whenever you have been hurt. Call one of the friends on your list if you need to talk about a problem.

Doing Nice Things for Yourself

To be good to yourself, you need to spend some time alone with yourself. Ask yourself every day how you are doing. If you feel sad, use self-talk to cheer yourself up. If you feel good, say something nice about yourself. Give yourself a compliment.

Taking care of your health is an important way to be good to yourself. Eating a healthy diet and exercising every day are two things you can do. Practicing good hygiene will help you feel good. Getting enough rest will also help keep you healthy. Remember that you are responsible for meeting your own needs. Taking care of yourself is an important part of having a friendship with yourself.

LESSON 1 REVIEW Write the answers to these questions on a separate sheet of paper. Use complete sentences.

1) How do friends help you meet your needs?
2) What are two things you can do to be a friend to yourself?
3) What are three ways you can make your self-talk positive?
4) What are two nice things you can do for yourself?
5) Suppose you and your sister try out for the soccer team. Your sister makes the team. You do not. How can you use positive self-talk to help yourself feel better?

Lesson 2

Making Friends

The first step in making friends is to be a good friend to yourself. Your friendship with yourself needs to be strong. Then you will be able to reach out to others. In this lesson, you will learn how to make friends.

What Kinds of Friends Do You Need?

Think about the kinds of people who make you feel comfortable. Do you like people who are loud or people who are quiet? Do you like being with large groups? Would you rather do things with only one or two people? Feeling relaxed and comfortable with friends is important.

Consider the kind of personality traits you look up to. You might say that a sense of humor is important. Loyalty and trust might be the most important things to you. You will want to choose friends who have traits you respect.

Also think about the part of your life in which you most need friends. Do you want to make more friends at school? Do you want to make more friends in your neighborhood? Maybe you want to make friends who enjoy the same sports you do. Friends usually have things in common. They often like the same things. They often enjoy the same activities.

Importance of Values

You have thought about the kinds of people you want for friends. You also need to think about your own values. Choosing friends who have the same values as you is important. For example, you know that you want to be drug free. Choosing friends who stay away from drugs would be best for you.

Keep in mind what you believe is right and good. For example, you may feel strongly about treating all people equally. You would not want to join a group that makes fun of others. By standing up for your own values, you will find lasting friends.

Fitness Tip

Exercising is a good way to spend time with friends. In this way, you do something together that you enjoy. You also do something that is good for your health.

What are some values that you and your friends have in common?

To find friends who like sports, you can go to a local park.

Forming Friendships

Once you know the kinds of friends you need, you can look for friends. You can let others know you are interested in being friends. Suppose you would like to meet new people in your neighborhood. You especially want to find friends who like sports. You might begin by going to a local park or gym. There you can meet people who also like sports. If you act friendly to people, chances are they'll be friendly to you, too.

Small talk
Talk about things that are interesting but not important

You may begin by making **small talk** with people. Small talk is talk about things that are interesting but not very important. You might ask for help or give a compliment. You might point out something you noticed. You might ask if you can join in an activity. Remember to be positive. If you practice what to say ahead of time, you will feel calmer. Your pleasant manner will lead people to want to know you better.

You can let the other person know you are interested in him or her. Being a good listener is important. A good listener pays attention to what the other person is saying. Looking at the person is a sign that you are interested. Ask questions from time to time. Let the speaker finish before you talk. You also might lean a little bit toward the person who is speaking. If you smile, the other person will feel comfortable.

Writing About Health

Think about what you want to find out when you get to know a person. List the things you think are important.

Once you feel relaxed talking with a person, take the next step in friendship. Ask the person to spend time with you. Sharing activities gives you a chance to learn about each other.

Careers

RESIDENTIAL COUNSELOR

Residential counselors help small groups of people handle their problems. The counselors may live with these groups in halfway houses. Halfway houses help people who are recovering from substance abuse. The counselors may also spend time with people in crisis shelters. People who are abused get help in crisis shelters. Residential counselors give emotional support to people. They help the people learn how to live on their own. Counselors supervise other people who work at the houses or shelters. A residential counselor must have training and be certified. Usually a counselor needs to have two to four years of college.

If you begin a friendship, remember to treat your new friend nicely. You can say things that help your friend feel good. You might thank your friend for coming over. You might say "I had a great time." You might thank your friend for listening. You could say "Thanks for understanding how I feel."

Also respect your friend's rights. Your friend has a right to ideas that are different from yours. You will get along better if you do not force your ideas on others. A true friend does not pressure a person to act in a certain way. If you are a friend, you will respect the decisions the other person makes.

LESSON 2 REVIEW Write the answers to these questions on a separate sheet of paper. Use complete sentences.

1) What are three things you need to look for in a friend?
2) How can standing up for your beliefs help you find lasting friends?
3) How can you meet people who have the same interests as you?
4) What are two ways to be a good listener?
5) How can you be a good friend to another person?

Lesson 3

Healthy Relationships

> **Relationship**
> *A connection between people*

*E*veryone has a need to belong with people. To meet that need, people have **relationships**. A relationship is a connection to another person. You have relationships with people in your family. You have other relationships, such as those with friends. Having healthy relationships can help keep you happy.

What Makes a Relationship Healthy?

People in healthy relationships act in certain ways. They accept each other's differences. They also respect each other's values. A person in a healthy relationship thinks about the rights of the other person. The needs of the other person are important.

Differences Between People

Usually you become friends with people who have something in common with you. You may enjoy doing the same things. Yet, each person has his or her own life.

Perhaps you want to go somewhere alone without inviting your friend. Time alone is important in a healthy relationship. Every person needs time to do things on his or her own. Every person grows and changes as time goes on. Healthy relationships allow people to be themselves.

Your Rights, Needs, and Values

Your rights, needs, and values are important. A healthy relationship will respect these parts of you. You need to make sure your rights, needs, and values are being respected. Asking yourself the following questions can help you:

Your rights
- Can I say what I think to my friend?
- Does my friend like me for who I am instead of wishing I were different?
- Can I make my own decisions without getting pressure from my friend?

Your Needs
- Do I feel comfortable with my friend?
- Do I give and receive equally in this friendship?
- Do my friend and I have fun safely?
- Do my friend and I show that we appreciate each other?

Your Values
- Do my friend and I share many of the same values?
- Do my friend and I respect our own and each other's values?
- Do we each choose to stand up for our values?
- Does my friend support my values and beliefs? Do I support my friend's values and beliefs?

If you can answer "yes" to most of these questions, your friendship is healthy. If you have some "no" answers, you need to think about your friendship. Maybe you can talk with your friend about these questions.

Make friends with people who enjoy the same sports you do.

CHANGES IN FAMILIES

A Finnish scientist, Alexander Westermarck, suggested that people lived in family groups in prehistoric times. Over the years, family life has changed. For example, years ago, children, parents, grandparents, and other family members usually lived in the same place. In modern times, family members often move far away from one another.

Family members can still remain close. Modern technology has made it easier for people to keep in touch with one another. People talk to each other over the phone. They can keep in touch by using computers. Today family members can remain close even if they live far away from one another.

Shyness
Feelings of discomfort around others

What Can Make Friendship Difficult?

Sometimes making and keeping friends is not easy. One thing that makes it hard for some people to make friends is **shyness**. People who are shy feel uncomfortable around others. They may feel nervous. They may not try to meet new friends. Other people might think a shy person is stuck-up.

Many people feel shy from time to time. If you feel shy much of the time, you can overcome the shyness. You might work with someone you trust to deal with the shyness. You can practice ways to talk to others.

Remember that most people like to talk about themselves. Asking people questions about themselves shows your interest in them. For example, you might ask someone how long he or she has lived in the area. Asking for other people's ideas makes them feel important. You might ask questions such as "What do you think of the new math teacher?" Giving compliments to others helps them feel good. You might say such things such as "You are a good soccer player. Maybe you could give me some tips." When people answer your questions, they will probably start to feel comfortable around you. Then they might want to get to know you better.

Gossip
The spreading of rumors, usually untrue, about people

A problem in keeping friends is **gossip**. Gossip is the spreading of rumors or ideas about people. Rumors are not always true. Usually, rumors are about people who are not around to hear what is being said. Gossip can be hurtful. It takes away people's trust. Staying away from gossip shows that you appreciate your friends. Its also shows that you value honesty.

What Can You Do When a Relationship Doesn't Work?

Sometimes, friendships end. A painful time usually follows. You might feel sad or lonely. You might feel that you failed. The best way to get through this time is to be a friend to yourself. Remember the friendship you have with yourself will last a lifetime. Here are some ways to be your own best friend:

- Talk honestly with yourself. Think about your feelings. Remind yourself that your feelings are normal. Use self-talk to say nice things to yourself.

- Appreciate yourself. Do something you like to do. Give yourself a special treat. Remember to tell yourself what a good person you are.

- Take care of yourself. Eat well, get plenty of rest, and exercise.

You might also want to talk to a caring person about how you feel. Talking to someone is especially important if you are unhappy for several weeks. It also is important if your feelings are very strong. Getting help when you need it is another way of being a good friend to yourself.

What do you think your friends like about you?

LESSON 3 REVIEW Write the answers to these questions on a separate sheet of paper. Use complete sentences.

1) What is one sign of a healthy relationship?

2) What are two rights you have in a friendship?

3) What are two things that can make it hard to make and keep friends?

4) How can you get through a painful time when a friendship ends?

5) What should you do if you feel that your friend does not respect your values?

CHAPTER SUMMARY

- Friendship is a special bond between people.

- Before you can be a friend to others, you need to be a friend to yourself.

- To be a friend to yourself, you can learn to use positive self-talk. You can also do nice things for yourself every day.

- Spending time alone with yourself and eating well, exercising, and getting enough rest are important.

- Before making friends, a person needs to decide what kinds of friends he or she needs.

- Choosing friends who share your values and have common interests is important.

- Showing interest in others, asking questions, and listening to others can help a person make friends.

- Showing friends that you appreciate them and respecting a friend's rights are important.

- Healthy relationships are important to a person's happiness.

- In healthy relationships, friends allow each other to be themselves. A friend respects the other person's rights, needs, and values.

- Friends often can talk about their problems and work them out.

- Asking people questions about themselves shows your interest in them. It also helps them feel comfortable.

- Gossip is often hurtful to others. Spreading gossip can damage a friendship.

- People usually feel sad and lonely when a friendship ends.

- Remembering to be your own best friend helps you when a friendship ends.

Chapter 6 Review

Comprehension: Identifying Facts

On a separate sheet of paper, write the correct word or words from the Word Bank to complete each sentence.

WORD BANK	
bond	self-talk
exercising	shy
gossip	small talk
needs	time
relationship	values

1) Making _____ is one way to start a conversation with another person.

2) _____ usually happens when a person who is being talked about is not there.

3) Friends who have a healthy _____ respect each other's values.

4) People who are _____ often feel nervous or uncomfortable around others.

5) Friends have a special _____ , or feeling of closeness with each other.

6) Positive _____ helps you be a good friend to yourself.

7) Choosing friends who have the same _____ as you is important.

8) _____ is something healthy you can do for yourself every day.

9) Your values and _____ are important in a friendship.

10) You need to spend _____ with yourself.

Comprehension: Understanding Main Ideas

Write the answers to the following questions on a separate sheet of paper. Use complete sentences.

11) What is an example of a way you can change a put-down to positive self-talk?
12) How can spending time alone be a good thing to do for yourself?
13) What kind of personality traits should you look for in a friend?
14) Why is it important to choose friends who have the same values that you have?
15) How should you act if one of your friends makes a mistake?
16) How can you show one of your friends that you appreciate him or her?
17) What might it mean if your friend gets angry when you do things with other people?
18) What should you do if you have very sad feelings that last for several weeks?

Critical Thinking: Write Your Opinion

19) Why is it difficult for people to accept others who are different? What can you say to a friend who gets angry because your ideas are different?
20) How do you feel when another person does not seem to be listening to you? What could you do to get better at listening to others?

Test Taking Tip — When studying for a test, use a highlighter to mark things you want to remember. To review your notes, look at the highlighted words.

Deciding for Yourself

Being a Good Listener

Being a good listener is important in relationships with others. When you listen to both the words and feelings in what someone says, it helps you understand the person. For example, your friend might be very upset and say that he or she "hates" someone. By listening carefully and asking questions, you might be able to help your friend calm down.

Good listening begins with body language that shows interest. Looking directly at someone shows your interest. So does your facial expression and posture. Paying attention to body language, facial expression, and tone of voice helps you hear the message.

Sometimes people start thinking about how they will answer a person. When people are thinking about their answer, they do not listen carefully to what is said. After listening to someone, take time to think about your answer. When you answer, speak calmly. Let the person you are talking to know you care about what is happening. It may not be possible for you to solve another person's problem. You might be able to suggest where the person might get some help.

Observe what you do when people are talking with you. Think about whether there are things you want to do to become a better listener.

Questions

1) How can watching someone's body language help you understand what the person is saying?

2) How do you feel when you are talking to people who cross their arms or frown at you?

3) What can you say to a family member who does not pay attention when you are talking?

4) What can you say to someone when you don't understand what is said?

Unit Summary

- Emotions are automatic responses to thoughts or feelings.

- Acting without thinking can have bad results. Thinking first helps a person act reasonably.

- Understanding the reasons for emotions can help a person control actions.

- Behaving in a reasonable way helps people feel good about themselves.

- Positive ways to handle emotions include talking about them or doing physical activity.

- Frustration may occur when someone is blocked from achieving a goal. Frustration can cause anger or aggression.

- Coping with frustration involves finding what is blocking the goal. Then strategies can be created to get around the block.

- Stress can be good or bad. It gives you energy to accomplish goals.

- Too much stress can be harmful. Keeping physically healthy helps a person manage stress.

- Peer pressure is the influence other people your own age have on you. Peer pressure can be positive or negative. To resist peer pressure, ask questions, express your feelings, or say "no."

- Anorexia and bulimia are eating disorders caused by emotional problems. People with eating disorders cannot control the way they eat. A person with an eating disorder may need a doctor's care.

- Before you can be a friend to others, you need to be a friend to yourself.

- To be a friend to yourself, use positive self-talk, do nice things for yourself, and spend some time alone.

- Choosing friends who share your values is important.

- To make friends, show interest in others. Ask questions and listen to what other people say.

Unit 2 Review

Comprehension: Identifying Facts

On a separate sheet of paper, write the correct word or words from the Word Bank to complete each sentence.

WORD BANK		
anxiety	exercise	self-esteem
blocked	grief	self-talk
cope	peer pressure	stress
depressed	respect	values

1) The mixture of painful emotions due to loss is called _____.

2) Physical or emotional pressure can cause _____.

3) A person who feels _____ is sad and has low energy.

4) Understanding emotions can help you _____ in a positive way.

5) A frustrated person feels _____ from meeting goals.

6) How a person feels about himself or herself is called _____ .

7) A feeling like fear for which the reasons are not clear is _____ .

8) People your own age can use _____ to influence your decisions and actions.

9) Physical _____ can help you cope with stress and anxiety.

10) It is important to choose friends who have the same _____ .

11) Use positive _____ to be a good friend to yourself.

12) In a healthy relationship, people _____ each other's needs and values.

Comprehension: Understanding Main Ideas

Write the answers to these questions on a separate sheet of paper. Use complete sentences.

13) Name at least four kinds of emotions.
14) What kinds of physical changes can be part of emotions?
15) How can stress be either positive or negative?
16) What are some ways of coping with stress?
17) What are some positive ways to respond to peer pressure?
18) How can you be a good friend to yourself?

Critical Thinking: Write Your Opinion

19) Why do some people respond in negative ways to emotions?
20) Why do some relationships end badly?

"We are indeed much more than we eat, but what we eat can nevertheless help us be much more than what we are."
—Adele Davis, *Let's Get Well*

Unit 3

Nutrition

What are your favorite foods? How do you decide what to eat? How does the food you eat affect your health? Eating can be a pleasure. If you know which foods to eat, eating can also keep your body healthy. Eating well means eating more than just the foods you like. It also means choosing foods that meet your body's needs.

Making wise food choices isn't hard. In this unit, you will learn what affects your food choices. You will also learn why it is important to make wise food choices.

▶ Chapter 7 Diet and Health
▶ Chapter 8 Making Healthy Food Choices

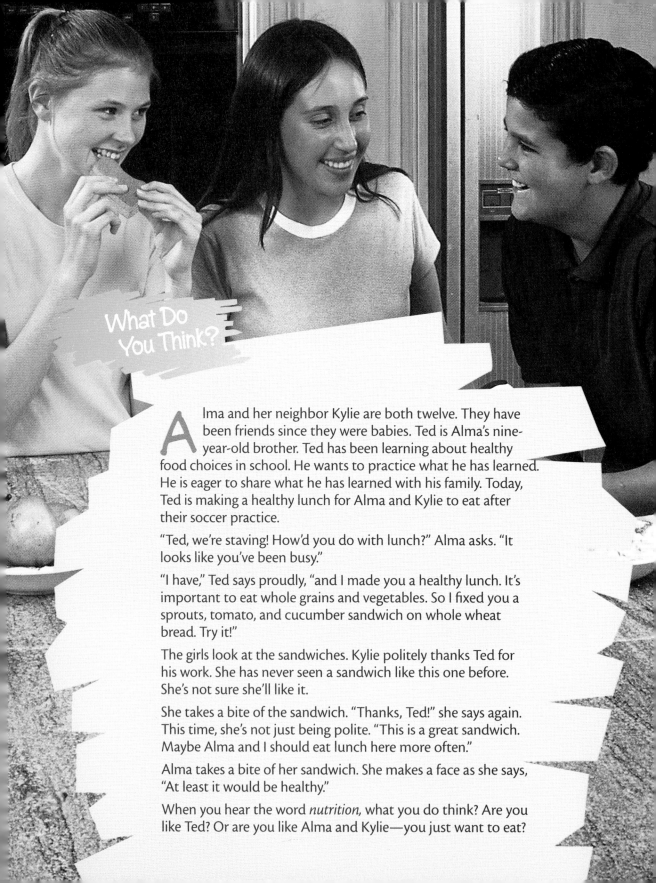

What Do You Think?

Alma and her neighbor Kylie are both twelve. They have been friends since they were babies. Ted is Alma's nine-year-old brother. Ted has been learning about healthy food choices in school. He wants to practice what he has learned. He is eager to share what he has learned with his family. Today, Ted is making a healthy lunch for Alma and Kylie to eat after their soccer practice.

"Ted, we're staving! How'd you do with lunch?" Alma asks. "It looks like you've been busy."

"I have," Ted says proudly, "and I made you a healthy lunch. It's important to eat whole grains and vegetables. So I fixed you a sprouts, tomato, and cucumber sandwich on whole wheat bread. Try it!"

The girls look at the sandwiches. Kylie politely thanks Ted for his work. She has never seen a sandwich like this one before. She's not sure she'll like it.

She takes a bite of the sandwich. "Thanks, Ted!" she says again. This time, she's not just being polite. "This is a great sandwich. Maybe Alma and I should eat lunch here more often."

Alma takes a bite of her sandwich. She makes a face as she says, "At least it would be healthy."

When you hear the word *nutrition*, what you do think? Are you like Ted? Or are you like Alma and Kylie—you just want to eat?

Chapter 7

Diet and Health

As you learned in Chapter 1, food is the body's fuel. Food has the nutrients that all the cells of your body need. However, not all food is equally healthy.

In this chapter, you will learn how food provides energy for the body. You will also learn what kinds of nutrients food can give the body. You will learn which foods contain these different nutrients. Finally, you will learn about dietary guidelines that can help you make healthy choices about the foods you eat.

Goals for Learning

▶ To describe how the body uses food for energy

▶ To explain that food provides calories and nutrients to the body

▶ To explain the importance of carbohydrates, proteins, fats, vitamins, and minerals in a healthy diet

▶ To identify healthy dietary guidelines

▶ To use the Food Guide Pyramid to judge food choices

Lesson 1

Food for Energy

Absorption
The moving of nutrients from the digestive system to the circulatory system

Digestion
The breaking down of food into nutrients

Metabolism
The process of cells using nutrients for energy and other needs

Nutrient
The basic unit of food that the body can use

The food you eat must be broken down into tiny parts. Only then can your body use it as energy. There are three main steps to turning the food you eat into energy. These steps are called **digestion**, **absorption**, and **metabolism**.

What Happens During Digestion?

During digestion, food is broken down into basic units that your body can use. These basic units of food are called **nutrients**. Digestion begins in your mouth. Chewing and chemicals in your saliva begin to break the food down. Digestion continues in the stomach. Once food is broken down, it moves into the small intestine.

What Happens During Absorption?

During absorption, nutrients are moved from the walls of the small intestine into the bloodstream. The circulatory system then carries the nutrients to all the cells in your body. If absorption did not occur, your cells would not receive the nutrients they need to live.

You need to eat healthy food so that your body can do all the things you like to do.

140 *Chapter 7 Diet and Health*

FOOD TECHNOLOGIST

Eating right can be a challenge. Food technologists work for universities, industries that make food or process food, and government agencies that make the guidelines for proper food storage and handling. To do this, a food technologist uses his or her knowledge of chemistry, microbiology, and other sciences to develop new ways to preserve, process, package, and store food. Usually a four-year college degree or higher is needed.

Calorie
The unit used to measure the amount of energy in foods

What Happens During Metabolism?

Metabolism occurs inside your cells. During metabolism, nutrients are taken into the cells. Inside the cells, nutrients are changed so the cells can use them for energy and other needs. As nutrients are used, wastes form. The cells remove these wastes.

What Are Calories?

You have learned that the body uses nutrients for energy. **Calories** measure how much energy different foods give the body. The more calories a food has, the more energy it gives the body. Extra calories are stored in the body as fat. Some foods have calories but few nutrients. Other foods have nutrients but few calories.

Fitness Tip

Go for a run or swim to burn off extra calories.

LESSON 1 REVIEW Write the answers to these questions on a separate sheet of paper. Use complete sentences.

1) List the three steps of using food for energy.
2) What happens during metabolism?
3) Where do the following steps occur in the body: digestion, absorption, and metabolism?
4) What do calories measure?
5) What might happen if you ate a diet high in calories but low in nutrients? Would this be healthy?

Lesson 2

Carbohydrates and Proteins

Carbohydrate
A nutrient needed mostly for energy

Protein
A nutrient needed for growth and repair of body tissues

Six nutrients are essential to life, but your body cannot make them on its own. Therefore, the foods you eat must contain these six essential nutrients for your body to be healthy. Two essential nutrients you need are **carbohydrates** and **proteins**.

What Are Carbohydrates?

Look at the foods in the picture below. Do you ever eat any of these foods? If so, you have been eating carbohydrates. When you eat foods with carbohydrates, your body breaks down the carbohydrates into a sugar. Your body uses this sugar as its main fuel. Extra sugar is stored in your liver for use later. Sometimes it is stored as fat. When you don't eat for a long time, your body can use the stored sugar for fuel. More than half of the foods you eat each day should contain carbohydrates.

What Kinds of Carbohydrates Are There?

There are two main kinds of carbohydrates: simple carbohydrates and complex carbohydrates. Table sugar, candies, cookies, and cakes all contain simple carbohydrates. These foods are sometimes called "empty calories" because they contain sugars and calories but few other nutrients.

Carbohydrates give the body much of the daily energy it needs.

142 Chapter 7 Diet and Health

Food	Total Carbohydrates (in grams)	Simple Carbohydrates (in grams)	Complex Carbohydrates (in grams)
Bread, 1 slice	13	1	12
Corn flakes, 1 oz. (low sugar)	24	2	22
Pasta or rice, ½ cup, cooked	20	0	20

Fiber
The parts of food that the body cannot digest

Complex carbohydrates are found in fruits, vegetables, grains, and many other foods. The foods in the table above are good sources of complex carbohydrates. Complex carbohydrates also have **fiber**. Fiber is not a nutrient, but it is important to your digestion and to your health. Fiber is the part of food that your body cannot digest. It helps move foods through the digestive system. Fiber is found in fruits, vegetables, grains, and beans.

What Are Proteins?

The picture below shows foods that contain proteins. Proteins are important for building muscles and repairing tissues. Proteins also provide energy.

First, your body uses the calories in proteins to build and repair tissues. Then, it stores the extra calories as body fat. About 12 to 15 percent of your diet should come from proteins each day.

Some protein foods are fish, meat, eggs, milk, poultry, peanuts, seeds, peas, and beans.

Amino acid
The smaller units of protein

Complete protein
A protein that has all nine essential amino acids

What Is a Complete Protein?

Proteins are made up of smaller units called **amino acids**. There are twenty-two amino acids. They join together to form all the different proteins that we eat. Your body can make some amino acids on its own. However, your body cannot make nine amino acids. You must eat these amino acids for your body to get them. A protein that has all nine of these amino acids is a **complete protein**.

Only foods that come from animal sources have complete proteins. These foods include meat, chicken, fish, eggs, milk, and milk products. Foods that come from only one plant do not have complete proteins. However, you can eat several different foods from plants at the same time to get complete proteins. For example, eating rice with beans gives you complete proteins.

LESSON 2 REVIEW Write the answers to these questions on a separate sheet of paper. Use complete sentences.

1) Compare simple carbohydrates with complex carbohydrates.

2) List five foods that are a good source of complex carbohydrates.

3) List five foods that contain proteins.

4) What is a complete protein? How can you get complete proteins?

5) Write down a meal you had today. List the carbohydrates and proteins in that meal.

Nutrition Tip

Skipping meals can rob your brain of the sugar it needs. Eating at least three meals a day at regular times can help ensure that your brain has enough sugar.

Lesson 3

Fats and Cholesterol

Cholesterol
A waxy, fatlike substance found in animal products

Polyunsaturated fat
A fat that is found mostly in plant foods

Saturated fat
A fat that is found mostly in animal products

The picture below shows foods that are high in fats. Like carbohydrates, fats also provide energy. Fats are found in all the cells of your body. Fats help the body use vitamins.

How Much Fat Should You Eat?

You need only small amounts of fats. Too much fat can lead to heart disease and cancer. Too little fat can also be harmful. Less than 30 percent of the calories you eat each day should come from fats. Salad dressings, margarine, dips, and gravies all add extra fat to your foods.

What Kinds of Fats Are There?

There are two main kinds of fats: **saturated fats** and **polyunsaturated fats**. Saturated fats are found mostly in animal products. They are usually solid at room temperature. Polyunsaturated fats are usually liquid at room temperature. They come mostly from plants.

What Is Cholesterol?

Cholesterol is a waxy, fatlike substance that is found in every cell of the body. Cholesterol is found in most of the foods that are high is saturated fats. It is found *only* in foods that come from animals.

Fats provide the body with energy but should be eaten sparingly.

Diet and Health Chapter 7

Healthy Subjects: Geography

A study was done of 1,800 Greenland natives during a 25-year period. In that time, only three people in the group studied had heart attacks. There have been similar findings in Japan and other Asian countries. This rate seems to be due to the large amount of fish that people in those places eat.

The rate of heart attacks in North America is more than ten times higher than in these other countries. Many Americans eat more meat, which has "bad" cholesterol, which clogs blood vessels. The oils in fish, on the other hand, contain "good" cholesterol, which helps blood flow. Many experts recommend eating fish twice per week.

You have probably heard of cholesterol as something bad. Your body needs some cholesterol to stay healthy. Too much cholesterol, however, can build up in your blood vessels. This makes it hard for your blood to flow. To protect your health, watch the amount of cholesterol and fats you eat each day.

LESSON 3 REVIEW Write the answers to these questions on a separate sheet of paper. Use complete sentences.

1) How might eating no fat harm your health?

2) List foods that contain fats.

3) How can you tell a saturated fat from a polyunsaturated fat?

4) How might eating too much cholesterol harm your health?

5) What percent of your daily calories should come from fat?

Lesson 4

Vitamins, Minerals, and Water

Mineral
A nutrient from the earth that is needed to help the body use energy from other nutrients

Vitamin
A nutrient needed to help the body use energy from other nutrients

Not all nutrients give your body energy. Carbohydrates, proteins, and fats do give your body energy. The other three essential nutrients do not provide energy. They are **vitamins**, **minerals**, and water. These nutrients help your body use the energy in food. Compared with carbohydrates, proteins, and fats, your body needs small amounts of vitamins and minerals. Your body needs enough water to replace the amount it loses each day.

What Are Vitamins?

The picture below shows foods that are rich in vitamins.

Vitamins A, D, E, and K dissolve in fats. They are easily stored in the body. This means that you can go a day without eating some of these vitamins if you have to. Also, if you take supplements, you could get too much of these vitamins.

Vitamin C and the B vitamins dissolve in water. Because they do, they can be lost by too much washing of foods or soaking them in water. They also do not stay in your body for long. Thus, you need to eat foods with these vitamins every day.

The body needs foods high in vitamins for normal growth.

Diet and Health Chapter 7

The chart below shows some of the jobs of different vitamins. It tells you which foods have which vitamins. Because different foods contain different vitamins, you should eat a wide variety of foods every day.

Vitamin	Job in the body	Foods that have this vitamin
Vitamin A	Keeps the skin, hair, and eyes healthy	Milk, egg yolk, liver, carrots, spinach
B Vitamins		
Niacin	Protects the skin and nerves	Meat, cereal, whole wheat, milk, fish, legumes
Thiamin	Protects nervous system, aids digestion	Pork, whole grains, green beans, peanut butter
Riboflavin	Protects the body from disease	Milk, eggs, bread, meat (especially liver), green vegetables
Vitamin C	Helps form bones and teeth, and resist infections	Citrus fruits, tomatoes, potatoes
Vitamin D	Helps form strong bones and teeth	Milk, fish oils
Vitamin E	Good for cells	Vegetable oils, margarine
Vitamin K	Helps blood clot	Green vegetables, soybeans, bran

What Are Minerals?

The picture on the following page shows foods that are high in minerals. Minerals are substances that are formed in the earth. Like vitamins, minerals do not give the body energy. You need to eat a wide variety of foods to get all the vitamins your body needs. You also need to eat many different foods to get all the minerals your body needs.

Foods high in minerals are important for digestion and fluid balance.

Which Minerals Does the Body Need?

Calcium, iron, and sodium are important minerals. Calcium and iron are needed for growth. Calcium helps bones and teeth form and stay strong. It is found in milk, yogurt, cheese, ice cream, green leafy vegetables, and beans. Girls and boys need about the same amount of calcium. This keeps their skeleton strong and their bones growing properly. Iron is found in meats, beans, and eggs. Red blood cells need iron to carry oxygen to other cells.

FOODS IN AND OUT OF SEASON

Since the 1800s, experts have suggested we eat fruits and vegetables for a healthy diet. Gardeners and shoppers could eat fresh fruits and vegetables only during summer. These foods were dried and stored for winter use.

After 1900, canned foods became available. Since 1930, we have been able to buy frozen fruits and vegetables. Because of better transportation, more countries can sell fresh produce. Now we can buy apples from New Zealand, pineapples from Hawaii, and cabbages from China. Even when some fruits and vegetables are out of season in the United States, our stores import and sell them.

Diet and Health Chapter 7

Sodium is important for nerve cells to communicate with one another. It is the main ingredient in table salt. It is also used to preserve foods. Sodium is common in packaged foods, for example. It's usually easy to get enough or even too much sodium from a typical American diet. Too much sodium can worsen high blood pressure.

How Is Water Important?

You body is about 60 percent water by weight. Water is needed to move nutrients through the blood and keep your body the right temperature. You can get the water you need each day by drinking eight glassfuls and from juices and soups.

Drink at least eight glasses of water daily to replace what the body loses.

LESSON 4 REVIEW Write the answers to these questions on a separate sheet of paper. Use complete sentences.

1) How are vitamins and minerals different from carbohydrates?

2) List three vitamins and the jobs they do in the body.

3) List three minerals and the jobs they do in the body.

4) How is water important to a healthy diet?

5) Why is it important to eat a wide variety of foods every day?

Lesson 5

Dietary Guidelines

Food Guide Pyramid
A chart that can be used to choose a healthy diet

There are so many things to consider when deciding which foods you should eat. How do you know if you are eating a healthy diet and getting all the nutrients your body needs? One aid is the **Food Guide Pyramid**, shown in Figure 7.1.

What Is the Food Guide Pyramid?

The U.S. government made the Food Guide Pyramid to help people decide how much and what to eat. Notice that the foods are divided into six groups.

Breads and grains form the base of the pyramid. This shows that more of these foods should be eaten than the other foods. As you move up the pyramid, fewer servings of each kind of food are needed in a healthy diet. Fats and sweets form the top of the pyramid. This group you should avoid or eat only in very small amounts.

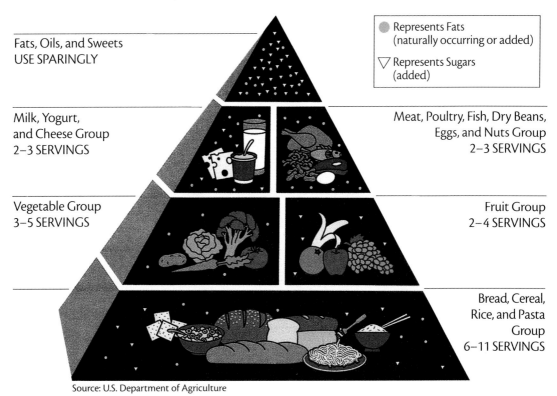

Source: U.S. Department of Agriculture

Figure 7.1. The Food Guide Pyramid

Serving size
A way to measure the amount of different foods that should be eaten each day

What Is a Serving Size?

Note that the Food Guide Pyramid suggests that you eat a certain number of *servings* from each food group each day. **Serving size** for each food group varies. The size of a serving depends on how many nutrients and calories the particular food has. For example, a serving in the fruit group might be a slice of melon, one medium banana, or $\frac{3}{4}$ cup of juice. For meat, which has more calories and fat than fruit, a serving size is 5–7 ounces. Each serving has about the same number of calories and major nutrients for each food.

Also, notice that the Food Guide Pyramid gives a range of servings. The smallest number of servings listed is the minimum amount of that group to be eaten each day. By eating the minimum amount, you'll still get the right amount of nutrients. Below is a list of more general guidelines to help you choose a healthy diet.

General Dietary Guidelines
- Eat a wide variety of foods.
- Balance the food you eat with physical activity.
- Eat many grains, fruits, and vegetables each day.
- Choose a diet low in fat, saturated fat, and cholesterol.
- Choose a diet low in sugar.

Action for Health

KEEPING A FOOD RECORD

Use the Food Guide Pyramid to plan a menu for yourself for next week. At the end of each day, check to see how many of the planned foods you actually ate. Monitor your energy level. Think about how you feel as you follow the suggested daily servings for each food group. Do you notice any difference in your ability to keep going without feeling tired or hungry between meals?

Writing About Health

Think about how your eating pattern compares with the suggestions in the Food Guide Pyramid. Write how you might change your eating pattern to make it more healthy.

Are Dietary Guidelines Different for Different People?

During different stages of your life, your body will need the same nutrients but in different amounts. Your body will also need different amounts of calories. It depends on your stage of growth or development.

For example, during pregnancy, a woman needs more calories, protein, calcium, iron, and some vitamins than usual. These extra nutrients keep her healthy and make sure the baby will grow properly. In a family with small children and teenagers, each person might have different dietary needs. Each person does not need to eat different foods. Each needs to eat the same foods in different amounts, depending on age, activity level, and gender.

technology

COMPUTER PROGRAMS FOR DIET MANAGEMENT

Did you know that computer programs can help people improve their food choices? Some programs use the Food Guide Pyramid to make sure your diet is healthy and well-balanced. Others use an index to show how different foods affect the amount of sugar in your bloodstream. This is especially useful in controlling obesity and diabetes.

Using a computer program, you can make a daily menu. The program may also suggest sports, activities, and exercises to improve your health. Quizzes rate how well you are eating.

Some computer games help make healthy eating habits. As healthy food shoots down junk food, health suggestions seep into your awareness. These computer programs also include a healthy eating guide, body weight logs, body mass index, and a calorie tracking system. Would you like to try one of these computer programs?

Healthy Subjects: Math

Have you ever wondered if your weight is right for your height and build? Find out by figuring out your body mass index. Let's say you weigh 145 pounds and are 6 feet tall.

1. Multiply your weight in pounds by 0.45.
   145 lb. × 0.45 = 65.25

2. Multiply your height in inches by 0.025.
   72 in. × 0.025 = 1.8

3. Multiply this number by itself.
   1.8 × 1.8 = 3.24

4. Divide the answer in step 1 by the answer in step 3.
   (65.25 ÷ 3.24 = 20.1).

Your number should fall between 19 and 25.

What Special Dietary Guidelines Do You Have?

Because you are growing and changing, you probably need more calories, calcium, and iron than your parents do. Boys are usually larger and have more muscle mass than girls. Boys usually need more calories than girls do. Girls who are menstruating need more iron than boys. That's because girls lose iron each month when menstruating.

LESSON 5 REVIEW Write the answers to these questions on a separate sheet of paper. Use complete sentences.

1) What is the Food Guide Pyramid?

2) List three general dietary guidelines.

3) What is serving size?

4) Give one way that dietary guidelines differ for girls and boys.

5) Why might you need more calories, calcium, and iron than your parents?

Chapter Summary

- The body uses three main steps to turn food you eat into energy. These steps are digestion, absorption, and metabolism.

- Digestion is the process of breaking down food into nutrients that the body can use.

- Absorption is the process of moving nutrients from the digestive system to the circulatory system. Once in the circulatory system, the nutrients are taken to cells throughout the body.

- Metabolism is the process by which cells use nutrients for energy and other jobs.

- Calories are used to measure the amount of energy in food.

- The body needs six essential nutrients. These are carbohydrates, proteins, fats, vitamins, minerals, and water.

- Carbohydrates, proteins, and fats give the body energy. Vitamins, minerals, and water do not give the body energy.

- Fiber is the part of food that the body cannot digest. Although fiber is not a nutrient, it is an important part of a healthy diet.

- Carbohydrates are found in grains, fruits, and vegetables. Complex carbohydrates give the body energy and fiber.

- Protein also gives the body energy. Proteins are important for building and repairing tissue. Meat, cheese, eggs, and beans are good sources of protein.

- A complete protein is one that contains the amino acids that the body cannot make. You can get complete proteins by eating foods from animals or by combining different foods from plants.

- The body needs fats. Saturated fats are less healthy than polyunsaturated fats. However, too much of any fat can lead to or worsen heart disease.

- The body also needs cholesterol. Too much cholesterol can build up in the blood vessels and prevent proper blood flow.

- Vitamins and minerals help the body get and use the energy in other nutrients.

- The Food Guide Pyramid can help you choose healthy foods in healthy amounts.

Chapter 7 Review

Comprehension: Identifying Facts

On a separate sheet of paper, write the correct word or words from the Word Bank to complete each sentence.

WORD BANK	
calories	polyunsaturated
digestion	metabolism
fiber	cholesterol
saturated	vitamins
Food Guide Pyramid	water
carbohydrates	absorption
amino acids	

1) Food is broken down into nutrients by a process called _____.

2) When nutrients move from the small intestine into the bloodstream, _____ is taking place.

3) The process of _____ takes place inside cells.

4) _____ are used to measure the amount of energy in different foods.

5) Simple _____ are sometimes called "empty calories."

6) The part of food that the body cannot digest is called _____.

7) A complete protein is one that contains all nine of the _____ that the body cannot make on its own.

8) Fats that are usually solid at room temperature are called _____.

9) _____ fats come mostly from plants.

10) Too much _____ can clog blood vessels.

11) _____ and minerals do not give the body energy.

12) You need to replace all the _____ that your body loses each day.

13) The _____ can help you determine how much and what kind of foods to eat.

Comprehension: Understanding Main Ideas

Write the answers to these questions on a separate sheet of paper. Use complete sentences.

14) What is a nutrient?

15) Give an example food for each of the six nutrients.

16) What does the body do with extra calories?

17) Give five dietary guidelines that most people should try to follow.

18) How can eating a wide variety of foods help you get the nutrients your body needs?

Critical Thinking: Write Your Opinion

19) Do you eat a healthy diet? Explain your answer using the Food Guide Pyramid.

20) Which of your favorite foods are healthy? List the foods and the nutrients they contain.

Test Taking Tip Always read test directions more than once. Underline the words that tell you how many examples or items you are to give. Check the directions again after you have finished the test to make sure you have not forgotten anything.

Chapter 8

Making Healthy Food Choices

The eating patterns you begin now will probably stay with you for the rest of your life. That's why now is a good time to learn about making wise food choices.

In this chapter, you will learn about healthy eating patterns. You will learn about the factors, such as advertising, that affect your food choices. You will also learn how to read a food label. Finally, you will learn how the government helps make sure foods are safe.

Goals for Learning

▶ To recognize healthy eating patterns
▶ To describe what influences your food choices
▶ To read a food label to make a healthy food choice
▶ To describe how the government makes sure our food is safe to eat

Lesson 1

Food Choices and Health

Your lifestyle and food choices affect your health. You face important food issues as you grow into a teenager. Some of these are your personal energy needs, snacking, weight changes, and diet-related problems. It's important to learn more about these issues and how they affect you.

How Much Energy Do You Need Each Day?

Your energy needs will vary during different stages of your life. Until you were one year old, you needed only 650–850 calories each day. Your energy needs increased each year as you grew older. Beginning at about age 11, energy needs begin to differ among boys and girls. Boys usually need more calories than girls starting at this time. Between the ages of 11 and 14, an average boy will need about 2,500 calories. An average girl of the same age will need about 2,200 calories. The number of calories recommended for a person is based on his or her height, weight, and activity level. Around the age of 50, both males and females need fewer calories. Pregnant women and women who are breast-feeding babies need about 500 extra calories each day.

Why Is Choosing Healthy Snacks Important?

Snacking is an important part of your diet. Your energy needs are high right now. This means that you need to eat more than you did just a few years ago. Snacks can help you get those extra calories and nutrients.

Snacks are also an important part of your lifestyle. You probably had some snacks with your family or friends over the last week. Look at the chart below. Did you choose any of these foods? For healthy snacks, choose these foods over cakes, candies, and high-fat cookies.

Nutrition Tip

Nutritious snacks can be part of a healthy diet. Be careful to avoid snacking in place of regular meals. Late night snacking can disturb your sleep and affect your weight.

Best Choices	Okay Choices
raw vegetables	flavored gelatin
fresh or dried fruit	fruited yogurt
sherbet or frozen yogurt	low-fat crackers or cookies

Make good choices when you choose snacks.

A study of teenagers' eating habits was conducted. It found that about one-third of a teen's diet comes from snacks rather than from meals. You can see then why choosing healthy snacks is important. Snacks are an important part of your diet. They also provide needed calories and nutrients.

How Can You Maintain a Healthy Weight?

Because adolescents grow at different rates, it is hard to give an exact number of calories. Body weight is either gained or lost depending on how many calories are taken in and how many are used by the body. As you grow into a young adult, your height and activity level will change. This will affect the number of calories your body needs and your weight.

Before you go on a diet, have a doctor or dietitian confirm that you need to change your weight. They can also help you make a healthy plan for changing your weight.

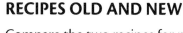

RECIPES OLD AND NEW

Compare the two recipes for plum pudding. One is from 1903. The other is from 1990.

Plum Pudding (1903)	Plum Pudding (1990)
3 cups beef fat	$1\frac{1}{2}$ cups vegetable oil
3 cups raisins	6 cups raisins and chopped fruit
3 cups water	4 cups whole wheat bread crumbs
1 tsp. salt	$1\frac{3}{4}$ cups fruit juice
$\frac{3}{4}$ cup sugar	1 cup brown sugar
3 cups flour	2 cups whole wheat flour
4 eggs	6 eggs
3 tsp. spices	3 tsp. spices
	3 tbsp. lemon rind

The new recipe has less fat and more vitamins from fruit. It also has fiber from whole wheat. Today, many recipes can be made healthier.

What Are Some Diet-Related Health Problems?

You know that eating too much or too little can affect your weight. It can also be a part of a larger eating disorder. An eating disorder is a pattern of eating that affects your health in a bad way. Eating disorders include anorexia and bulimia.

People who have anorexia choose not to eat because they think they are fat. A person with anorexia may eat little and exercise often until becoming life-threateningly thin.

People with bulimia tend to eat large amounts of food at one time. Then, they either throw up the food or use laxative drugs to rid their body of the food. A person with bulimia may appear to be the correct weight. However, the person can die from the unhealthy pattern of eating and ridding the body of food.

Bloated
Swollen

Malnutrition
A condition in which the body does not get enough to eat

Obesity
A condition in which one is more than 20 percent overweight

Obesity results when a person is more than 20 percent above a healthy weight. This condition often is caused by unhealthy eating habits and lack of activity. Obesity stresses the body and its organs. It can also cause related disorders.

Many people in the United States are overweight because they eat too much food. However, there are people around the world who do not have enough food to keep them healthy. Sometimes people even die from lack of enough food.

Malnutrition occurs when the body does not get enough calories or nutrients. There are two types of malnutrition. One type of malnutrition is caused by a lack of calories from protein. You may have seen pictures of children with this type of malnutrition. The children have thin legs and arms and huge, **bloated**, or swollen, stomachs. The second type of malnutrition occurs when a person does not get enough calories. This person is starving to death due to lack of food.

LESSON 1 REVIEW Write the answers to these questions on a separate sheet of paper. Use complete sentences.

1) What factors affect your energy needs?
2) Is snacking a part of a healthy diet? Explain.
3) Give three examples of healthy snacks.
4) Describe two eating disorders.
5) Explain why choosing healthy foods is important.

Lesson 2

Influences on Food Choices

Many things affect the way you eat. Usually, you probably are not even aware of all the reasons you eat certain foods. As a child, you ate foods that tasted good or that were prepared by people who cared for you. As you grew older, your friends and ads began to influence you.

How Do Feelings Affect Your Food Choices?

Sometimes you eat foods just because they help you feel good. For example, after a hard day at school, have you ever headed for the refrigerator to find something to eat? Eating is usually a good experience. It can mask feelings of anger or frustration. Sometimes, people eat food to avoid facing their feelings or problems.

Other times, food is used as a reward or a punishment. Have you ever been rewarded with a special meal when you had a good report card? Or have you ever been sent away from the table because you were misbehaving? In both cases, your feelings were probably affected.

DIETARY AIDE

Hospitals and nursing homes need dietary aides. These workers help prepare food for residents. Many dietary aides work on a tray assembly line. They follow directions about what should go on each person's tray. Then they deliver the trays. After a meal, they pick up the trays and return them to the kitchen. They record the amount of food patients eat. Aides may help sort recyclables and trash, wash dishes, and do other clean-up tasks. Dietary aides are sometimes called food service workers. Dietary aides usually receive on-the-job training.

Making Healthy Food Choices Chapter 8

Children choosing what to eat in the school cafeteria.

How Does Your Environment Affect Your Food Choices?

Your environment also affects your food choices. You may make most of your food choices in your home environment. You make choices based on which foods are available.

A restaurant is another environment that can affect your food choices. The foods you can choose are limited to those on the menu. A restaurant may offer only foods high in fat and calories. Then your healthy food choices will be limited.

You also make food choices at school. You may choose to eat in the school cafeteria or to bring your own lunch. At school, the foods available in the cafeteria line affect you. So do the food choices of your friends and peers.

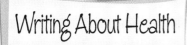

Writing About Health

Think about the food choices you have made. Compare one influence that resulted in a healthy meal with one that resulted in poor food choices.

164 Chapter 8 *Making Healthy Food Choices*

How Do People Affect Your Food Choices?

You might make some food choices for social reasons. For example, you might eat a food you don't like just because your friends eat it. Suppose a friend makes fun of a food you like to eat. Then you might decide not to eat it around that friend. You might decide that you don't really like the food after all.

As a baby, you couldn't make many food choices. Your family gave you foods that they thought were good for you. Your family probably still has a lot of influence on your food choices. Sometimes family members can influence you to eat foods that you don't really want. For example, after seeing your parents eat certain foods over time, you may be influenced to try those foods.

This family influence can be positive because you might try vegetables that you otherwise would not. But it can also be less positive. For example, you might find it hard to say no to sweets and desserts. You might not want to hurt someone's feelings.

How Does Culture Affect Your Food Choices?

Culture can have a lot of influence on your food choices. Many foods you eat with your family may be recipes passed down from generation to generation. You might also eat traditional foods on holidays that are special to your culture or religion. One benefit of learning about other cultures is that you can try foods that you might not be used to.

How Does Advertising Affect Your Food Choices?

Food companies spend millions of dollars studying why people buy certain foods. Using this information, they target different groups of people in their ads and food packages. When advertisers target young children, for example, they make their ads using cartoons or toys. These ads usually show how much fun the food is to eat. Adult ads might focus on nutrition or convenience rather than on fun.

Fitness Tip

Take a walk after a large meal.

Action for Health

EXAMINE FOOD ADVERTISING

For a few days, look for food and drink ads on TV and radio, in magazines and newspapers, and on billboards. Then walk through your local grocery store. Find two or three products that you saw ads for. Notice where in the store these items are located.

Then, carefully read the list of ingredients on the packages. The ingredients that are listed first will make up more of the food than the other ingredients. What are the main ingredients in the products? How healthy do you think these products are? Do you think the advertising that you saw presented a fair picture of the product?

A food's package is also made to attract a person's attention. Often the food is put in a large or colorful package. Then, it is easier to spot in the store. Foods are put into large displays in supermarkets to get your attention. Foods that small children like are put on low shelves that they can reach.

LESSON 2 REVIEW Write the answers to these questions on a separate sheet of paper. Use complete sentences.

1) List four things that affect your food choices.
2) Describe a time when your feelings affected your food choices.
3) What kinds of food choices are available in your school cafeteria?
4) What is the purpose of a large, colorful food display in a grocery store?
5) Make a list of the reasons you choose certain foods during one day.

Lesson 3

Reading Food Labels

Lot number
A number that identifies a group of packages

An important part of healthy eating is being able to read and understand the labels on packages. By reading food labels, you can determine the nutritional value of foods. You can also compare one product with another one, so you can make the healthier choice.

What Does the Package Tell You?

In the United States, food sold in packages must have certain information on the label. The package must include what is inside and how it should be stored. The package should give the weight and **lot number** of the food. A lot number is a number that identifies a group of packages. Many packages have a certain date by which the food should be used.

Some things that must be on the package are listed below:

- the name of the product
- the product's weight
- a list of the product's ingredients
- the manufacturer's name and address
- a food label

Healthy Subjects
Math

In the past, scientists used the calorie to measure energy. However, the popularity and widespread use of the International System of Units changed the unit scientists use to measure energy. The International System of Units is usually abbreviated SI, for *le Système International d'Unités*. It is sometimes referred to as metric units.

The SI unit for measuring energy is the joule. Because scientists still have to use the old research, they sometimes have to convert back and forth from calories to joules.
One calorie is 4.184 joules.

CHUNK LIGHT TUNA IN WATER

Nutrition Facts	Amount/Serving	%DV*	Amount/Serving	%DV*
Serv. Size 2 oz. drained (56g / about ¼ cup) Servings about 2.5 **Calories** 60 Fat Cal. 5	Total Fat 0.5g	1%	Total Carb. 0g	0%
	Sat. Fat 0g	0%	Fiber 0g	0%
	Cholest. 30mg	10%	Sugars 0g	
	Sodium 250mg	10%	**Protein** 13g	23%

*Percent Daily Values (DV) are based on a 2,000 calorie diet.

Vitamin A 0% • Vitamin C 0% • Calcium 0% • Iron 2%
Niacin 20% • Vitamin B-6 8% • Vitamin B-12 20% • Phosphorus 8%

INGREDIENTS: LIGHT TUNA, WATER, VEGETABLE BROTH, HYDROLYZED CASEIN, HYDROLYZED SOY PROTEIN, SALT.

Figure 8.1. What can you find out by reading this food label?

> **Nutrition Facts**
> *The part of a food label that tells about the calories and nutrients in the food*

What Are Food Labels?

You have probably seen many food labels. Figure 8.1 above shows a food label from a can of tuna. Food labels are one piece of information that must be included on a food's package. The United States government sets guidelines that tell which information must be on a food label.

What Information Can You Get From a Food Label?

Find the heading **Nutrition Facts** on the food label above. The Nutrition Facts tell the size of one serving. Note that one serving of tuna is 2 oz., or about $\frac{1}{4}$ cup. The label also tells how many servings are in the whole package. If you eat the whole can of tuna, you will get 2.5 servings. Nutrition Facts also list the number of calories per serving and the number of calories from fat. For each serving of the tuna that you eat, you will get sixty calories. Five of these sixty calories will be from fat.

Daily Values
The part of the food label that tells about the percent of nutrients in the food, based on 2,000 calories per day

What Are Daily Values?

Another part of the Nutrition Facts is the percent of the **Daily Value** for each nutrient in the food. These percentages tell you how much of that nutrient is found in the food. These numbers are based on a diet of 2,000 calories per day. Find the Daily Value percentage for sodium on the food label in Figure 8.1. One serving of tuna has 10 percent of the Daily Value for sodium.

Under Daily Values, a food label must give the amounts of total fat, cholesterol, sodium, total carbohydrate, and protein. Amounts are given in either grams (g) or milligrams (mg). You learned in Chapter 7 that these are important nutrients to track in your diet. Thus, you can see how helpful food labels can be in choosing to eat a healthy diet.

What Other Information Is on the Food Label?

The Nutrition Facts must also show the amount of vitamin A, vitamin C, calcium, and iron in the food. What percentage of these vitamins and minerals are found in tuna? Notice that the label may list other vitamins and minerals. This information is also listed as a percentage of the 2,000 calories daily diet.

The food label must also give all the ingredients that are in the food. Find the list of ingredients on the food label in Figure 8.1. These ingredients are listed in order by weight, from the most to the least. For example, in the can of tuna, light tuna is the first ingredient. This means that the weight of the tuna weighs more than any other ingredient in the can.

Look at the food label for a can of soup. You will probably find that water is the first ingredient. The water in the soup probably weighs more than any other ingredient. The list of ingredients can help you find out whether the food has an ingredient you are allergic to. It's also helpful for people who need to avoid certain ingredients for other health reasons.

Action for Health

COMPARING FOOD LABELS

Collect food labels from several brands of a similar food. For example, you might collect the food labels for five different kinds of chips or cold cereals. Make sure the foods are similar.

Then, carefully read the food labels for each food. First, check the serving sizes. Are they the same? Why might a company alter its suggested serving size? Then compare the Nutrition Facts for each food. What does your comparison show? Is one brand healthier than another? Will this information change your food choices in the future?

Why Should You Read Food Labels?

All of the information on the food label can be helpful in making healthy food choices. If you are trying to limit a substance in your diet, the label tells you how much is in one serving of the food.

By using the information on the food label, you can tell if the food is priced fairly. You can look at the price and the number of servings and figure out how much each serving costs. Then, compare that amount with other brands to see which product is the best value.

LESSON 3 REVIEW Write the answers to these questions on a separate sheet of paper. Use complete sentences.

1) What three pieces of information can you find on a package of food?

2) What are Daily Values based on?

3) What can the order of ingredients on a food label tell you about the food inside?

4) Why is it important to know the serving size of a food?

5) Look at the food label for tuna in Figure 8.1. For which vitamins does tuna provide more than 10 percent Daily Values?

Lesson 4

Making Foods Safe

Additive
A chemical added to food to make it better in some way

Enrichment
Adding extra nutrients to a food

Fortification
Adding a nutrient that a food lacks

Preservative
A chemical added to food to prevent spoiling

Many government agencies monitor the safety of all food sold in the United States. These agencies make rules that help keep foods safe. They can also help if you feel that food safety is at risk in places that sell food. You can take steps in your own home to reduce the risk of illness from foods.

What Does the FDA Do?

You learned about food labels in Lesson 3. The U.S. Food and Drug Administration (FDA) determines what information to include on food labels. The FDA also determines which foods can be sold and what claims manufacturers can make about the foods they sell. An example of a nutritional claim is "low fat" or "high in fiber." Manufacturers must follow the guidelines set up by the FDA when they make such claims.

If you have read the list of ingredients on many food labels, you know that many chemicals are added to foods. The FDA makes sure that these chemicals are tested and are safe to eat.

Why Are Chemicals Added to Food?

Chemicals are added to foods for many reasons. For example, **additives** are added to a food to make it better. An additive may be added to a food to make it last longer. Additives also make foods taste better. They improve the texture and appearance of foods, too.

Preservatives are added to food to keep germs from growing in them. Preservatives allow food to travel long distances across the United States without spoiling. This allows more people to be fed.

Sometimes vitamins and minerals are added to foods for **enrichment** or **fortification**. Enrichment means nutrients are added to the food because they have been lost during processing. Breads and cereals are often enriched with minerals to make them healthier.

Contaminate
Infect by contact with germs or toxins

Fortification means certain nutrients are added because the food is naturally low in those nutrients. Milk is fortified with vitamin D. Some brands of orange juice are fortified with calcium. Sometimes foods that are low in all nutrients are fortified to make them healthier. Punch is often fortified with vitamins. Some sugared breakfast cereals are also fortified with nutrients. Without fortification, punch and some breakfast cereals would provide calories but few nutrients.

Which Other Agencies Control Food Safety?

The U.S. Department of Agriculture (USDA) and the Public Health Department are two other agencies that work to keep food safe. The USDA is a federal agency. The Public Health Department is a state or local agency.

The USDA grades foods for quality so that buyers know what they are buying. Meat is an example of a food that receives grades from the USDA. The USDA also makes sure that meat-packaging plants are clean. They carry out inspections to make sure that the foods are packaged and stored properly.

The Public Health Department inspects restaurants. They make sure that restaurants are clean. They make sure that safe practices are used when food is handled. A health inspector might check the temperature of refrigerators and steam tables at a restaurant. Inspectors want to make sure cold foods are kept cold enough and hot foods are kept hot enough.

Why Should You Learn to Handle Food Properly?

There are two main reasons why you should learn how to handle foods properly. First, you want to make sure nutrients are not lost when you store and cook foods. Second, you want to make sure that foods do not make you sick. Foods can become infected, or **contaminated** with germs or toxins that cause food poisoning. Handling foods properly prevents this.

Wash fruits and vegetables before eating or cooking them.

How Can You Keep Nutrients in Your Foods?

One way to keep nutrients from being lost is to use fresh foods within days after buying them. When preparing foods, cook vegetables as little as possible. Remember that most vitamins and some minerals dissolve in water. If foods are cooked in water, you may be throwing away vitamins and minerals with the water. You can steam vegetables or quickly stir-fry them instead.

How Can You Keep Foods Safe to Eat?

Cold foods should be kept cold—at 40 degrees Fahrenheit or below. Hot foods should be cooked until they reach 140 degrees Fahrenheit. Germs grow quickly at room temperature. Foods left at room temperature for even twenty minutes can spoil and can cause illness. Therefore, leftovers should be refrigerated right after you eat a meal.

Food can be contaminated with germs when people who are sick handle it. Do not handle foods if you have an open wound on your hands. You should also always wash your hands with warm, soapy water before handling any foods or utensils. Foods can be contaminated if the countertop or your utensils have germs on them. Always cook with clean utensils and in areas that have been cleaned properly.

Always clean up carefully and quickly after cooking with meat. If you put raw chicken on a cutting surface, clean the surface with hot, soapy water before putting any other ingredients on it. Otherwise, those ingredients could become contaminated with germs. Use the rules below to help you remember these safe practices.

Rules for Keeping Food Safe
- Keep cold foods cold and hot foods hot.
- Always wash your hands with warm, soapy water before touching food or utensils.
- Wash surfaces and utensils before using them.
- Do not handle foods if you have an open wound on your hands.

LESSON 4 REVIEW Write the answers to these questions on a separate sheet of paper. Use complete sentences.

1) What are the names of two agencies that help keep food in the United States safe?

2) What is the difference between fortified food and enriched food?

3) What are two things you can do to make sure the foods you prepare do not lose their nutrients?

4) What are three things you can do to keep the foods you cook at home safe?

5) You have just cut up raw chicken on a cutting board. What should you do next?

Chapter Summary

- The amount of energy you need will change over the course of your life. A person's energy needs depend on his or her height, weight, activity level, and stage of growth.

- Teenagers may get as much as one-third of their daily calories from snacks. Choosing healthy snacks is one way for teenagers to get the extra nutrients and calories that are needed for growth.

- Anorexia and bulimia are life-threatening eating disorders. By eating too little or too much, a person can harm his or her body.

- Malnutrition results from lack of calories or nutrients.

- Feelings, environment, people, culture, and advertising can all affect your food choices. It is important to know what these things are so you can make healthy food choices.

- If only unhealthy foods are available, you will have limited healthy choices. You may choose to eat some foods because others like them, rather than because you like them.

- Advertisers use different techniques to get people to buy certain foods. Foods are advertised in ways that will make you notice them and want to buy them.

- Food labels include information about the food inside. Labels include Nutrition Facts, Daily Values information, and a list of ingredients.

- Nutrition Facts tell serving size, number of servings in the package, and number of calories per serving.

- Daily Values show the percentage of nutrients in a 2,000-calorie diet.

- The U.S. government has set up agencies to make sure foods are labeled correctly and are safe to eat.

- Different chemicals are added to foods. Additives are used to improve flavor or some other characteristic of food. Preservatives are used to prevent spoiling.

- By handling foods safely, you can prevent food poisoning and keep foods nutritious.

Chapter 8 Review

Comprehension: Identifying Facts

On a separate sheet of paper, write the correct word or words from the Word Bank to complete each sentence.

WORD BANK	
fortification	serving size
Public Health Department	weight
diet	USDA
food poisoning	wash
chemicals	room temperature
nutrients	snacks
enrichment	

1) Adding calcium to orange juice is an example of _____.

2) Adding nutrients to punch is an example of _____.

3) Energy needs vary depending on your height, _____, activity level, and stage of growth.

4) _____ are an important part of a healthy diet.

5) Preservatives are _____ that are added to foods to prevent spoiling.

6) If foods are not handled properly, they can lose their _____.

7) When left at room temperature, some foods can become contaminated and can cause _____.

8) Daily Values are based on a 2,000-calorie _____ per day.

9) The _____ on a food label tells you how much of the food is used for the Nutrition Facts.

176   Chapter 8   *Making Healthy Food Choices*

10) The _____ inspects restaurants.
11) The _____ inspects meat-packaging plants and grades meat products.
12) Before handling food, you should always _____ your hands with warm, soapy water.
13) Germs can grow quickly at _____.

Comprehension: Understanding Main Ideas

Write the answers to these questions on a separate sheet of paper. Use complete sentences.

14) If you think that you are overweight, what should you do?
15) How might a food advertiser try to convince a young child that its product is good?
16) List three things that affect your food choices.
17) What are some reasons additives are added to foods?
18) List three things you can do to handle food properly.

Critical Thinking: Write Your Opinion

19) Why do you think the U.S. government has agencies to help keep our food safe?
20) Before you studied this chapter, did you regularly use safe food-handling practices? Explain your answer.

Test Taking Tip When you read test directions, try to restate them in your own words.

Words Used in Advertising Foods

Think about some of the words you see in food ads and on food packages. Have you ever seen the words *light, low fat, nonfat,* or *fat free*? These words are common. They seem to say that any fat in foods is bad.

Advertisers use these and other words to make us think that their foods are healthy. They want us to think that eating their foods will make us healthy. Many people think they are making healthy food choices when they buy these foods.

It is true that eating too much fat is not healthy. But the body needs some fat to function properly. It's important to read all the information on a food label. Don't just read the advertised words. For example, notice how many calories the food has. The amount of salt or sugar is also important.

The best way to stay healthy is to eat foods from all six food groups and to exercise regularly.

Questions

1) Look for food ads. You can find them in newspapers or magazines. Cut out the ads. What kind of ads did you find?

2) List the words that might make someone think the foods are healthy. Then write down how you feel when you hear these words.

3) List any negative words that hint that you could harm your health by choosing other foods.

4) Why is reading the food label the best way to tell how healthy a food is?

Unit Summary

- The body needs six essential nutrients. These are carbohydrates, proteins, fats, vitamins, minerals, and water.

- Carbohydrates, proteins, and fats give the body energy.

- Vitamins, minerals, and water help the body to use other nutrients and to function properly.

- Carbohydrates are found in grains, fruits, and vegetables.

- Protein gives the body energy. Proteins are important for building and repairing tissue. Meat, cheese, eggs, and beans are good sources of protein.

- The body needs fats. Saturated fats are less healthy than polyunsaturated fats. Too much fat can lead to heart disease.

- The body also needs cholesterol. Too much cholesterol can build up in the blood vessels and prevent proper blood flow.

- The Food Guide Pyramid can help you make healthy food choices.

- The amount of energy you need depends on your height, weight, activity level, and stage of growth.

- Feelings, environment, people, culture, and advertising can affect your food choices.

- Advertisers use different techniques to get people to buy foods.

- Food advertising is intended to affect your food choices. Foods are advertised in ways that will make you notice them and want to buy them.

- Food labels include Nutrition Facts, Daily Value information, and a list of ingredients. Food labels can help you know how healthy the food is.

- The U.S. government has set up agencies to make sure foods are labeled correctly and are safe to eat.

Unit 3 Review

Comprehension: Identifying Facts

On a separate sheet of paper, write the correct word or words from the Word Bank to complete each sentence.

WORD BANK		
calories	Daily Values	water
USDA	wash	vitamins
Food Guide Pyramid	carbohydrates	complete
	serving size	nutrients
weight	preservatives	

1) Your energy needs depend on your _____, height, activity level, and stage of growth.

2) _____ are added to foods to prevent spoiling.

3) If foods are cooked in lots of water, you may be throwing away _____ with the water.

4) _____ are based on a 2,000 calorie diet per day.

5) Before handling food, you should always _____ your hands with warm, soapy water.

6) _____ are used to measure the amount of energy in foods.

7) Simple _____ are sometimes called "empty calories."

8) A _____ protein is one that contains all nine amino acids that the body cannot make on its own.

9) You need to replace all the _____ that your body loses each day.

10) The _____ can help you determine how much and what kind of foods to eat.

11) _____ and minerals do not give the body energy.

12) The _____ inspects meat-packaging plants and grades meat products.

13) _____ is a way to measure the amount of different foods that you should eat.

Comprehension: Understanding Main Ideas

Write the answers to these questions on a separate sheet of paper. Use complete sentences.

14) What is the Food Guide Pyramid?

15) List the six essential nutrients you should eat each day.

16) What are three things you can do to handle food safely?

17) What are two pieces of information you can find on a food label?

18) Describe three things that affect your food choices every day.

Critical Thinking: Write Your Opinion

19) Why do you think it is important to make healthy food choices every day?

20) What can you do to improve your food choices?

Drug misuse is not a disease. It is a decision, like the decision to step out in front of a moving car. You would call that not a disease but an error of judgment.
—Phillip K. Dick, *A Scanner Darkly*

Unit 4

Use and Misuse of Substances

What do you think of when you hear the word *drug*? You might think of a medicine you took for a cold or other illness. Or, you might be reminded of a tragedy you have read about that resulted from using an illegal drug.

Medicines are drugs used to cure and treat disease. They are wonderful and important scientific tools that benefit people. In this unit, you will learn to use medicines safely. You will also learn the dangers of illegal drugs and of misusing legal drugs.

▶ Chapter 9 Medicines and Drugs

▶ Chapter 10 Drug Dependence—Problems and Solutions

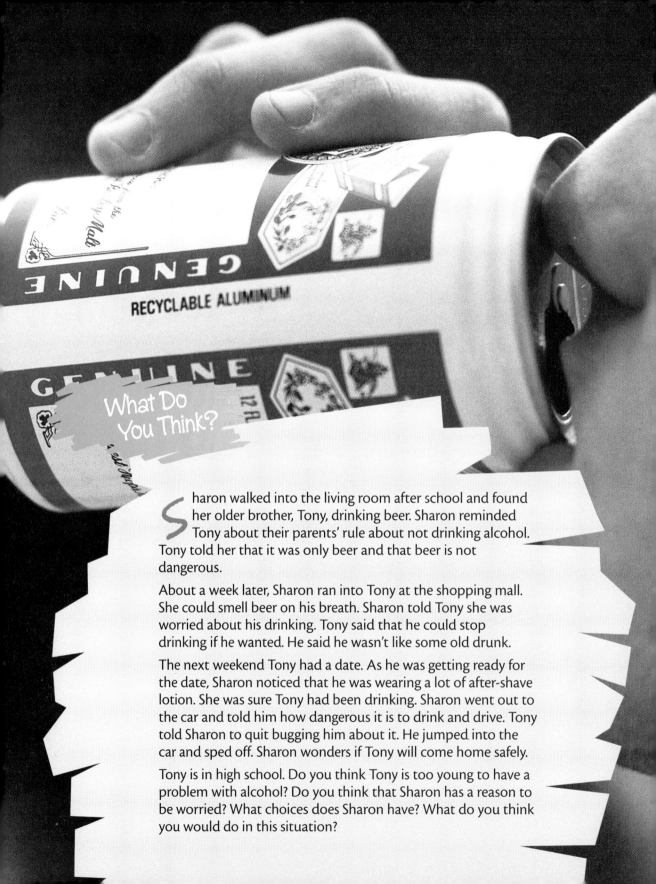

What Do You Think?

Sharon walked into the living room after school and found her older brother, Tony, drinking beer. Sharon reminded Tony about their parents' rule about not drinking alcohol. Tony told her that it was only beer and that beer is not dangerous.

About a week later, Sharon ran into Tony at the shopping mall. She could smell beer on his breath. Sharon told Tony she was worried about his drinking. Tony said that he could stop drinking if he wanted. He said he wasn't like some old drunk.

The next weekend Tony had a date. As he was getting ready for the date, Sharon noticed that he was wearing a lot of after-shave lotion. She was sure Tony had been drinking. Sharon went out to the car and told him how dangerous it is to drink and drive. Tony told Sharon to quit bugging him about it. He jumped into the car and sped off. Sharon wonders if Tony will come home safely.

Tony is in high school. Do you think Tony is too young to have a problem with alcohol? Do you think that Sharon has a reason to be worried? What choices does Sharon have? What do you think you would do in this situation?

Chapter 9

Medicines and Drugs

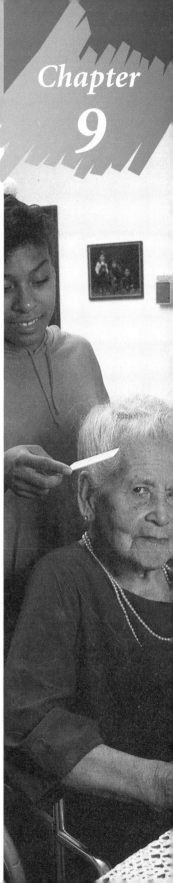

Have you recently taken a medicine? Did it help you feel better if you were sick? Did it help make you well? Did it help you avoid becoming sick? What do medicines do? How do they work? Why are there so many different kinds? How do they differ from drugs?

In this chapter, you will learn about medicines. You will learn how they can improve and save lives. You will also learn about harmful drugs.

Goals for Learning

▶ To define the terms *drug* and *medicine*

▶ To explain the difference between prescription medicines and over-the-counter medicines

▶ To describe how medicines and drugs affect the body

▶ To identify the health risks associated with tobacco use

▶ To describe the effects that alcohol has on the brain and the body

▶ To list some often-abused drugs and describe dangers associated with each

Lesson 1

Prescription and Over-the-Counter Medicines

Drug
A substance that changes the way the mind or body works

Medicine
A drug used to treat or prevent a disease or health problem

Pharmacist
A person trained and licensed to prepare and sell prescription drugs

Prescription
A written order from a doctor for a medicine

Over-the-counter medicine
A medicine that can be bought without a doctor's written order

People can use **drugs** or **medicines** to relieve pain or restore health. A drug is a substance that changes the way the body or mind works. A medical drug, also called a medicine, is used to treat or prevent disease, or to relieve discomfort. Medicines can be classified by how they are bought.

What Are Prescription Medicines?
Some medicines are sold by **prescription** only. You must have a written order from a doctor to buy a prescription medicine. A **pharmacist** can fill a prescription. A pharmacist is a trained and licensed professional who fills prescription medicines.

What Are Over-the-Counter Medicines?
Some medicines are sold without a doctor's prescription. These are called **over-the-counter medicines**. You probably are familiar with many kinds of over-the-counter medicines. There are medicines for minor pain and fever, for colds, and for minor skin problems. There are so many similar over-the-counter medicines that it is hard to choose among them. It is a good idea to let your doctor or pharmacist suggest one for your needs.

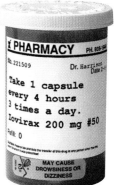

Read the labels on prescription drugs carefully.

What Do Different Medicines Do?
Thousands of different prescription and over-the-counter medicines are available in the United States. These medicines can be grouped by the purpose for which they are used.

Pain Relievers
Mild pain relievers are available over the counter. You may have taken aspirin or other remedies for a headache or muscle pain. These medicines also reduce fever. For severe pain, doctors prescribe stronger pain relievers. Strong pain relievers are sold by prescription only. A doctor must supervise their use because they can be dangerous in large doses.

Over-the-counter medicines are usually not as strong as prescription medicines. Why do you think this is so?

Action for Health

BEING PREPARED FOR DRUG REACTIONS

There is some risk in using any kind of drug or medicine. Occasionally, people have a dangerous reaction to a drug. They may be allergic to the drug, take too much of it, or combine it with other drugs. A serious drug reaction can cause permanent damage to a person's health.

You can be prepared for a drug reaction. Think about what you would do in a drug reaction emergency. Talk with your family members and friends to make a list of actions you could take. Find telephone numbers for a poison control center and the nearest hospital emergency room. Keep these numbers next to your telephone in case a drug reaction emergency should ever happen in your home.

Antibiotic
A drug used to fight bacterial infections

Antihistamine
A medicine used for treating allergy symptoms

Decongestant
A medicine that opens lung and nasal passages

Ointment
A medicine for minor skin infections

Antibiotics

Antibiotics are medicines that stop the growth of germs that cause disease. You might have taken an antibiotic for a sore throat or other infection. Most antibiotics are prescription medicines. Antibiotics, however, do not stop the growth of all germs. Antibiotics do not affect germs that cause colds and flu. The only antibiotics sold over the counter are found in **ointments**, or medicines for minor skin infections.

Symptom Relievers

Many medicines relieve the symptoms of diseases. For example, **decongestants** open lung and nasal passages. **Antihistamines**, medicines used for treating allergy symptoms, dry up a runny nose. Cough medicine quiets coughing. Some of these medicines are sold over the counter. Other stronger medicines of this kind are prescription medicines.

Consult an adult before using over-the-counter medicines.

Medicines and Drugs Chapter 9 **187**

Antibody
A protein that is stimulated by the immune system to fight disease

Cardiovascular
Relating to the heart and blood vessels

Cardiovascular medicine
A drug used for the heart and blood vessels

Diabetes
A disease in which the body does not make enough insulin

Psychoactive medicine
A medicine that changes the function of the brain

Vaccine
A medicine that stimulates the immune system to fight off a disease

Are all medicines drugs? Why or why not?

It is important that symptom relievers, as well as pain medicines, are not overused. Pain or a cough might be a symptom of a serious disease. Relieving the symptoms does not cure the disease. Rather, it may mask the disease. All symptoms are important clues to a doctor. Your doctor may have diagnosed the cause of your symptoms. In that case, he or she may allow you to use symptom relievers to be more comfortable.

Cardiovascular Medicines

Cardiovascular diseases affect the heart and blood vessels. These diseases are major health problems for adults in this country. Some **cardiovascular medicines** improve the function of the heart and blood vessels. Others lower blood pressure or reduce fats in the blood. These medicines are sold by prescription only. One over-the-counter medicine, however, may help the heart. Aspirin may help prevent heart attacks in some adults.

Chemical Replacements

Some diseases are caused by the lack of a certain chemical in the body. An example is the disease **diabetes**. The body cannot make enough of a chemical, insulin, that controls the amount of sugar in the blood. People with this disease get insulin by giving themselves a shot. This helps make their blood sugar level normal. Some people can control the disease by changing their diet and taking medication.

Vaccines

Some medicines prevent disease. A **vaccine** is a prescription medicine that can prevent diseases, such as measles, mumps, and chicken pox. The vaccine stimulates the immune system to produce **antibodies** to a possible disease. Antibodies are proteins that fight diseases. Vaccines are prescription medicines that you can get in a doctor's office or a health clinic.

Psychoactive Medicines

Some medicines help the brain function normally. A medicine that affects thoughts and emotions is called a **psychoactive medicine**. Some psychoactive medicines help people with depression, and others reduce anxiety.

Food and Drug Administration (FDA)
A government agency that oversees the testing and sale of medicines

All psychoactive medicines are prescription medicines. A doctor must supervise the use of psychoactive medicines.

How Do We Know Medicines Are Safe?

In the United States, the **Food and Drug Administration (FDA)** oversees the testing and sale of medicines. New medicines must go through many tests before they can be sold. Medicines are carefully tested before the FDA approves their use. The FDA determines whether a medicine will be sold over the counter or by prescription. The FDA also tells the maker what the medicine's label should say. Labels must state the purpose of the medicine, the recommended dosage, possible side effects, and other cautions. Always read and follow the directions on medicines. If you need help reading a label, ask an adult for assistance.

LESSON 1 REVIEW Write the answers to these questions on a separate sheet of paper. Use complete sentences.

1) What is the difference between a medicine and a drug?
2) What is the difference between an over-the-counter drug and a prescription drug?
3) Describe what a pharmacist does.
4) List four purposes of medicines.
5) How do you know if it is safe for you to take a medicine?

PATENT MEDICINES

Around 1900, newspapers were full of ads for "patent medicines." These medicines were said to cure one or more illnesses or problems. But the medicines were often fifty percent or more alcohol. Manufacturers never tested them for safety or the accuracy of the claims. People wanted to feel better. So they believed the claims and ordered the medicines. Sellers became rich, but the medicines didn't help the buyers. Today the Food and Drug Administration (FDA) regulates medicines. Researchers must prove a medicine does what it claims. No medicine can be sold without FDA approval.

Lesson 2

The Effect of Medicines and Drugs on the Body

Inject
Use a needle to take medicine into the body

Suppository
A cylinder containing medicine to be inserted into the rectum

To affect the body, a medicine must be taken into the body and enter the bloodstream. After a drug enters the bloodstream, it is carried to all parts of the body.

How Are Medicines Taken?

Medicines can be taken in several different ways. You probably have taken medicine by mouth. The way a medicine is given is based on the nature of the medicine.

- Some medicines can be taken by mouth, or orally—in liquid, pill, or capsule form.

- Some medicines are **injected**—either under the skin, into a muscle, or into a vein. Injection is the fastest way to get medicine into the body. Most injections use a metal needle that makes a hole in the skin.

Do medicines that are taken orally begin to work right away? Why or why not?

- Some medicines can be absorbed through the skin. The body slowly absorbs medicine from a special patch. Some heart medicines and antismoking medicines are given this way.

- Some medicines are given in a **suppository**. A suppository is inserted into the rectum. The medicine is absorbed through the thin mucous membranes.

PHARMACY CLERK

Pharmacy clerks usually work in hospitals or drug stores. Under a pharmacist's direction, they measure out or mix prescriptions. They deliver prescriptions to nursing units in the hospital. They may be responsible for figuring out the cost of medicines. They also check the pharmacy's stock of medicines. They order and keep records of medicines and supplies. They also may type and file reports and answer phones.

Most states require pharmacy clerks to be licensed. A clerk may take a five- to ten-month program or a two-year program at a community college.

Codeine
A prescription medicine that relieves severe pain

Side effect
An unexpected and often harmful result of taking medicine

Therapeutic effect
A helpful result of taking medicine

How Does a Medicine Affect the Body?

When a medicine has spread throughout the body, it begins to affect certain parts of the body. A **therapeutic effect** is the helpful effect the medicine provides. For example, **codeine** is a prescription medicine. Its therapeutic effect is relief of severe pain. But codeine, like some medicines, has unwanted effects, called **side effects**. Side effects can be as minor as an uncomfortable feeling or a dry mouth. Some side effects are dangerous or can be life threatening. A doctor must weigh the therapeutic effects against the side effects. For some patients, some side effects are too severe to deal with. In this case, the doctor must try a different medicine.

Mixing one or more medicines can cause problems. For example, a person who takes one medicine may also need to take another medicine. Sometimes two different medicines cause a bad chemical reaction in the body.

A person can become dependent on a medicine. For example, someone might take a medicine to help sleep. If the person takes the medicine too frequently, he or she may not be able to sleep without it. This is one reason it is important not to take more of a medicine than your doctor prescribes.

Why Do Medicines Affect Different People Differently?

Many things influence how a medicine affects a person. A person's sex, age, weight, or amount of body fat all affect how medicines act in the body. A doctor considers these things when he or she decides the dose you need. A heavy adult male may need more of the same medicine than a thin, young female.

Another important thing that influences how a medicine affects a person is body chemistry. Some kinds of medicine have possible side effects. One person might have no side effects from a medicine. Another person might have serious side effects. Side effects are difficult to predict. It is important to ask your doctor or pharmacist about side effects for each medicine you take.

Health Tip

Some drugs have side effects. Discuss with your doctor alternatives to treatment with medicines.

An unborn baby is much smaller than the woman. Any drug has a greater effect on the baby than on the woman.

How Can Taking Medicines Affect an Unborn Baby?

A pregnant woman must be careful about the medicines she takes. Many medicines can enter her unborn baby's body. Many medicines that are safe for adults are harmful for a developing baby. Unborn babies who are exposed to medicines or drugs can be smaller than normal. They may have damaged internal organs or other permanent disabilities. These children may need special care throughout their lives.

A woman who is pregnant or who might be pregnant should ask her doctor before she takes any medicine. Even over-the-counter medicines can harm unborn babies. Medicines and drugs most easily hurt a developing baby during early pregnancy. Of course, a pregnant woman's use of tobacco, alcohol, or any drug is dangerous for her baby. Some unborn babies are exposed to medicines and drugs before their mothers know they are pregnant. Any woman who might be pregnant should understand the risks of medicine use.

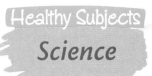

PREVENTING DISEASE AND BIRTH DEFECTS

Scientists working in biological and genetic sciences continue to find better ways to prevent disease and birth defects in unborn babies. Women who are pregnant can now be tested early enough in their pregnancy to detect the possibility of a serious disease.

What Should You Think About When Taking a Medicine?

Medicines contain powerful chemicals. Medicine can help you if you are sick. If used in the wrong way, however, medicine can make you sick. It is important to follow safety rules for taking medicine.

- Let your doctor know about all of the medicines that you take, even vitamins.
- Carefully follow the directions for taking the medicine. Ask your doctor if you should take all of the prescription medicine in a container.
- Read and follow all warnings on the medicine container. For example, some medicines might make you sleepy. A label might warn you not to drive or use machinery.
- Tell your doctor immediately about any side effects.
- For both prescription and over-the-counter medicine, take only medicine given to you by a trusted adult. *Never* accept medicine from anyone else.
- Store medicines properly. Do not take prescription medicine unless it was prescribed for your current illness.
- Do not mix medicines unless your doctor has told you to do so. *Never* mix medicines and alcohol.

Writing About Health

Some people take medicines that aren't necessary and may be harmful. Write why you think it is important to take only necessary medicines.

LESSON 2 REVIEW Write the answers to these questions on a separate sheet of paper. Use complete sentences.

1) What are two ways drugs are taken into the body?
2) Why do some drugs have bad side effects?
3) Why can two people react differently to the same drug?
4) Why must a pregnant woman be careful about taking drugs?
5) Write a list of five safety tips for using drugs wisely.

Lesson 3

Tobacco

Nicotine
A chemical in tobacco to which people become addicted

Stimulant
A drug that speeds up the central nervous system

A drug is a chemical substance other than food that changes the way the mind or body works. The main reason people continue to smoke is because of **nicotine**, a chemical in tobacco. Tobacco use is legal for adults over 18 years of age. Using tobacco, however, has been proved to be harmful to a person's health. Today, the majority of people in the United States do not smoke or use other tobacco products. Many people who used to smoke have quit as they learned the possible harm that tobacco can do.

Cigarettes, cigars, chewing tobacco, pipe tobacco, and snuff all contain tobacco. Many people who use tobacco began when they were teens. Choosing to start using tobacco may not seem harmful at the time. But tobacco users quickly become dependent on the powerful drug in tobacco.

Why Do People Use Tobacco?

Many older people began to smoke before the harmful effects of tobacco were known. They may believe that smoking helps them relax or concentrate better.

Why do some young people think smoking tobacco makes them seem older?

Younger people may begin to smoke cigarettes because of peer pressure. Smoking may make them feel more comfortable in social settings.

What Does Tobacco Do to the Body?

Tobacco has many harmful effects on the body. Tobacco use is linked to cancer, cardiovascular disease, and lung disease. A 30-year-old person who smokes two packages of cigarettes a day might shorten his or her life by about eight years.

How can tobacco shorten someone's life? As mentioned above, tobacco contains the chemical nicotine. Nicotine is a **stimulant**. Nicotine raises the heart rate and blood pressure. It narrows blood vessels. This reduces the amount of blood that reaches some parts of the body. These actions are hard on the heart and can cause someone's early death.

Emphysema
A serious disease of the lungs that causes difficulty in breathing

Lung cancer
A disease of the lungs caused primarily by smoking tobacco

Secondhand smoke
Tobacco smoke breathed by nonsmokers

Smokeless tobacco
Tobacco that is chewed

Tar
A substance in tobacco that can form a thick, brown, sticky substance in the lungs

Cigar smokers do not inhale. Does this make cigar smoking a safe thing to do?

What Are the Harmful Effects of Cigarette Smoke?

Burning tobacco produces more than two thousand harmful chemicals. These include **tars**. Tars are dark-colored, sticky substances. Tars contain smoke particles and many different chemicals. Tars damage the interior surface of lung passages. A lung damaged by tars cannot clean itself. This can lead to a serious lung disease called **emphysema**. People with emphysema have difficulty breathing. They cannot push air out of their lungs.

Lung cancer, one of the most deadly cancers, is linked to smoking. Smoking causes more than 90 percent of lung cancer cases. Most people who get lung cancer will die from it. But lung cancer is not the only cancer linked to smoking. Smokers have high rates of mouth and throat cancer, bladder cancer, and cancer of the pancreas.

How Does Smoking Affect Other People?

Being near a smoker can harm your health. **Secondhand smoke** is cigarette smoke that pollutes the air. Secondhand smoke can cause disease in nonsmokers. Breathing secondhand smoke causes the heart to beat faster. It also causes blood pressure to rise.

Adults who smoke may put their families at risk. Children of smokers have more respiratory illnesses than children of nonsmokers. Babies of smoking mothers are born smaller than other babies. These babies are more likely to die before birth or in the first months of their life.

Are There Safe Forms of Tobacco?

There is no safe way to use tobacco. Some people smoke cigarettes that have less tar and nicotine than regular ones. These people, however, may smoke more often. They may inhale smoke more deeply. They probably do this to get the same amount of nicotine as from a regular cigarette.

Some people use **smokeless tobacco**. Smokeless tobacco is put between the cheeks and gums and chewed. In the past, people believed that it was safe to use smokeless tobacco.

Withdrawal
A physical reaction to the absence of a drug

Although users of smokeless tobacco do not get lung damage from smoke, they absorb many poisons from tobacco. They become dependent on nicotine, as smokers do. The use of smokeless tobacco can cause the gums to pull away from the teeth so that the teeth become loose and fall out. Smokeless tobacco can also cause cancer in the mouth. Using smokeless tobacco is not safer than smoking tobacco. It is not safe at all.

How Can People Stop Using Tobacco?

Many people who use tobacco realize that they should quit. They are addicted, however, to nicotine. People who try to stop using tobacco may go through a period of **withdrawal**. This is a physical reaction to the absence of a drug in the body. They may have headaches, be unable to sleep, or have a hard time concentrating. People who choose to stop using tobacco can get help. They can get skin patches or special gum. These patches and gum have a small amount of nicotine. This helps a person slowly stop using nicotine.

What Are the Rights of Nonsmokers?

People have begun to learn how harmful secondhand smoke can be. As they have, steps have been taken to protect the health of nonsmokers. First, restaurants began to have nonsmoking sections. Many restaurants now do not allow smoking at all. You may have seen signs on restaurants telling customers the restaurant is smoke free.

All tobacco products contain nicotine and can cause health problems.

People used to be able to smoke on airplane flights. The smoke stayed in the cabin during the flight so that everyone had to breathe it. Now all airplane flights within the United States are nonsmoking. This rule protects both the passengers and the people who work for the airlines.

What do the warnings printed on tobacco products say?

Because of the harm that secondhand smoke can do, many laws now forbid smoking in public places. These laws protect the rights of nonsmokers. Today there are fewer places where it is legal to smoke. This may help people decide that it is worth the effort for them to quit.

How Can You Avoid Secondhand Smoke?

Breathing cigarette smoke is dangerous to your health. Nonsmokers can get some of the same physical effects that smokers get. Stay away from areas in which people are smoking. Look for no-smoking signs in restaurants and other buildings. Better yet, try to stay in areas that are posted as smoke-free environments. In most places today, it is easy to avoid breathing cigarette smoke.

LESSON 3 REVIEW Write the answers to these questions on a separate sheet of paper. Use complete sentences.

1) Why do people become addicted to smoking tobacco?
2) What is smokeless tobacco?
3) What diseases can be caused by smoking tobacco?
4) How does smoking tobacco endanger the health of nonsmokers?
5) Why are smoke-free buildings good for people's health?

Lesson 4

Alcohol

Alcohol
A chemical that depresses the central nervous system

Alcoholic beverage
A drink that contains alcohol

Depressant
A drug that slows down the central nervous system

Disinfectant
A chemical used to prevent the spread of disease

Alcohol is a substance that has many uses. You might have used alcohol to clean a computer keyboard. Alcohol works better than other cleaners because it evaporates so quickly. Alcohol is also used as a **disinfectant**, or a chemical to prevent the spread of disease. Many medicines, both prescription and over-the-counter, contain alcohol. You might have taken a medicine for a cold that contained alcohol as one of the ingredients.

But the most common use of alcohol is as a drug. Drinks that are partly alcohol include wine, beer, and liquors. Any drink that is part alcohol is called an **alcoholic beverage**.

Drinking alcoholic beverages is legal for adults over 21 years of age. In some states, it is legal for people over 18 years of age to drink beer or other liquor. There are reasons, however, why some people should never drink alcohol at all. No one should abuse alcohol.

Why Do People Use Alcohol?

People use alcohol to feel comfortable in social situations, to give themselves more self-confidence, or to relax. Some people use alcohol for the wrong reasons. They may be trying to escape uncomfortable emotions. Using alcohol for this reason may have a bad effect. The alcohol may temporarily cover up the bad emotions, but it doesn't change them. Then the effects of the alcohol wear off. The emotions—and the problems that are causing them—are still there.

Many teens may try alcohol because of peer pressure. This is not a good reason to use alcohol. First, alcohol is probably illegal for teens in your state. And, alcohol use by teens can cause serious physical, mental, and social problems.

What Does Alcohol Do to the Body?

Alcohol is a central nervous system **depressant**. It slows the activity of the brain and decreases muscle control and coordination. When people use alcohol, they will not be able

Alcohol abuse
Drinking too much alcohol or drinking too frequently

Intoxicated
Excited or stimulated by a drug

to hear, see, or think as well as they ordinarily would. The effects of using alcohol happen quickly after the drug is used. The more alcohol that is taken, the greater the effects will be.

Alcohol is taken into the stomach. Then it is absorbed into the bloodstream and taken to various parts of the body. Too much alcohol can be harmful. The body cannot absorb the drug quickly enough, so the person becomes **intoxicated**, or drunk.

What Happens If Too Much Alcohol Is Used?

An intoxicated person experiences major changes in body functions. These include confused thinking, loss of motor control, difficulty walking, and slurred speech. Alcohol is a psychoactive drug. It affects the mind or mental processes and can interfere with a person's ability to think. A person who drinks too much alcohol may become emotional or violent.

Extreme amounts of alcohol can make you sick. Large amounts can cause unconsciousness and sometimes death. Drinking too much alcohol, drinking alcohol too frequently, or combining alcohol with other drugs is called **alcohol abuse**.

What Are the Harmful Effects of Alcohol?

Drinking too much alcohol over a long period of time can permanently damage the body. The brain, heart, kidneys, stomach, intestines, and liver can all be harmed by alcohol abuse.

When Is It Dangerous to Use Alcohol?

It is always dangerous to drink too much alcohol. Combining alcohol with medicine can be unsafe also. A person should never use alcohol before driving a car or using dangerous machinery such as an electric drill.

How Does Society Encourage the Safe Use of Alcohol?

Every year hundreds of people are killed in car accidents caused by someone driving while intoxicated. Many of the people killed are children or teenagers—innocent victims

All of these drinks contain about the same amount of pure alcohol.

Alcoholism
A disease in which a person is dependent on the use of alcohol

Designated driver
The person in a group who will not drink alcohol and will drive the group home

Ethyl alcohol
A kind of alcohol found in beer, wine, and hard liquors

of alcohol abuse. Many people are working to change this situation. Two such groups are Mothers Against Drunk Driving (MADD) and Students Against Drunk Driving (SADD). There may be groups such as these in your community.

One idea that can help reduce deaths from drunk drivers is the **designated driver**. In a group that will be drinking, the designated driver is chosen ahead of time. He or she drives the group home. The designated driver does not drink. Choosing a designated driver protects the lives of the people drinking as well as the lives of people in other cars.

What Is Alcoholism?

People can become both physically and psychologically addicted to **ethyl alcohol**. Ethyl alcohol is found in beer, wine, and hard liquors. **Alcoholism** is a severe mental and physical dependence on alcohol. People with the disease of alcoholism cannot control their drinking. They can hurt themselves and their families and lose their jobs or quit school. They may also begin having major health problems.

technology

SOFTWARE FOR ALCOHOL TESTING

People who drink and drive are dangerous to themselves and others. Breath analyzers help police keep drunk drivers off the roads. These analyzers measure blood alcohol levels. One analyzer is called "ignition interlock." It is put on the cars of people who have had problems with drinking and driving. The device is wired to the car's starter. Before starting the car, the driver blows into the device. Sensors check for alcohol in the breath. If they sense alcohol, the car will not start. The computer records the number of failed tests. It can even tell if the system has been tampered with. Such technology keeps everyone safer.

Drunk drivers kill hundreds of people each year. Students Against Drunk Driving (SADD) works to stop people from driving after drinking.

Alcoholics Anonymous (AA)
An organization that helps people live alcohol-free lives

How Can People With Alcoholism Get Help?

The disease of alcoholism has no cure. People with alcoholism must stop drinking and never drink again. Many community mental health centers offer alcohol treatment programs at low cost. A nationwide group helping people with alcoholism is **Alcoholics Anonymous (AA)**. AA groups help people lead alcohol-free lives.

Are Some Kinds of Alcoholic Drinks Safer Than Others?

Some people believe that drinking wine or beer is safer than using "hard liquors" such as whiskey, vodka, or gin. This is not true. One serving of each type of alcoholic beverage contains about the same amount of alcohol. Drinking too much of any kind of alcohol can damage the body.

What are some ways to recognize a driver who has been drinking?

LESSON 4 REVIEW Write the answers to these questions on a separate sheet of paper. Use complete sentences.

1) Name three uses of alcohol.

2) How does alcohol affect the body?

3) What are some health effects caused by drinking too much alcohol?

4) Why is it dangerous to drive a car after using alcohol?

5) What can people in your community do to prevent or reduce alcohol abuse?

Lesson 5

Narcotics, Depressants, Stimulants, and Hallucinogens

Hallucinogen
A drug that confuses the way the brain processes information

Heroin
A dangerous and illegal narcotic drug

Narcotic
A drug that dulls the senses or relieves pain

Opiate
A drug made from opium poppy plants; another name for narcotic

Synthetic
A narcotic drug that is manufactured in laboratories

Tolerance
A condition in which a person must take more and more of a drug to get the same effect

You might have heard the term *narcotic* used to describe all illegal drugs. As you will learn, **narcotics** are just one group of often-abused drugs. Three other groups are depressants, stimulants, and **hallucinogens**.

What Are Narcotics?

Narcotics are a group of drugs that have a true medical use. Morphine and codeine are used for severe pain, such as pain after surgery or in late stages of cancer. Narcotics are also called **opiates** because many of them are produced from the opium poppy plant. Other narcotics are **synthetic**. They are manufactured in laboratories.

Narcotic drugs can cause unpleasant side effects. They slow breathing, cause sleepiness, confuse thinking, and cause nervousness. No one should ever take narcotic drugs without the advice and supervision of a doctor.

Narcotics can produce a powerful physical and mental dependence. So, these drugs are very habit forming. A person taking the drug builds up a **tolerance** to it. More and more of the drug is needed to get the same result. As the person takes more of the drug, an overdose may occur. This could cause unconsciousness or even death.

Heroin is a dangerous and illegal narcotic drug. Heroin has no legal or medical use in the United States. Any person who buys or sells heroin is breaking the law. Heroin is usually injected into a vein with a needle. This means that heroin users also risk diseases, such as AIDS, from using contaminated needles.

What Are Depressants?

A depressant is a drug that slows the activity of the brain. You have read about alcohol, which is a depressant. Tranquilizers, barbiturates, and nonbarbiturate sleeping drugs are also

Convulsion
A drawing tightly together and relaxing of a muscle

Tremor
Severe shaking

depressant drugs. Most depressants are prescription medicines. Doctors may prescribe tranquilizers for anxiety or barbiturates to help people sleep.

Depressants produce a calming effect because they slow down the central nervous system. They lower blood pressure and heart rate. Using a depressant drug can cause sleepiness or confused thinking. It may be dangerous to drive a car or operate machinery while taking depressant drugs.

Depressants are often abused. A user can quickly become dependent on the drug. Then, more and more of the drug must be taken to get the same result. Sometimes a person cannot get a doctor's prescription for more of the drug. Then he or she may begin buying it illegally. Because depressants affect the brain centers that control breathing, an overdose can cause respiratory failure.

Health Tip

Use physical exercise instead of drugs to relax your body and relieve anxiety or stress.

A person who is dependent on a depressant may not be able to stop using the drug suddenly. In users, the brain has changed physically. It has "learned" to operate with the drug. When a depressant is stopped suddenly, the user may have **tremors**. The person may shake severely or have his or her muscles drawn tight in **convulsions**. Suddenly stopping a depressant can be life threatening. A person who is trying to stop using alcohol or any other depressant should be under a doctor's care.

The table below lists some types of commonly abused drugs.

Narcotics (opiates)	Depressants	Stimulants	Hallucinogens
Codeine	Alcohol	Nicotine	LSD
Morphine	Tranquilizers	Amphetamines	PCP
Heroin	Barbiturates	Cocaine	MDMA (Ecstasy)
Synthetic opiates	Nonbarbiturate sleeping drugs	Crack cocaine	Mescaline
			Psilocybin

Why would a person misuse both depressants and stimulants?

Amphetamine
A synthetic stimulant

Caffeine
A stimulant found in coffee, tea, chocolate, and some soft drinks

Cocaine
A dangerous and illegal stimulant drug made from the coca plant

Crack cocaine
A form of cocaine that is smoked

Ice
A form of an amphetamine that is smoked

What Are Stimulants?

A stimulant is a drug that increases the activity of the brain. You have read about the stimulant nicotine. **Caffeine** is the mild stimulant found in coffee, tea, soft drinks, and chocolate.

The other stimulants in this group are dangerous drugs. **Cocaine** is an illegal drug made from the coca plant. It is a white powder that is sniffed through the nose. It can also be injected into a vein. **Crack cocaine** is a powerful form of cocaine that is smoked. Dependence on cocaine can occur rapidly—sometimes with the first use. People who use cocaine stop eating properly, and malnutrition may result.

Amphetamines are synthetic stimulants. Some amphetamines are prescription medicines. For example, a doctor might prescribe an amphetamine drug to help someone lose weight. But, doctors do not prescribe these drugs as often as they once did. There is too much of a risk that a person will become dependent on the drug. There are several different kinds of illegal amphetamines. **Ice** is a form of amphetamine that is smoked. Ice can cause permanent mental disability.

Stimulant drugs speed up the heart and breathing. At first, they cause a user to feel clearheaded and not tired. Later, users feel nervous, anxious, and irritable. A user may become confused, fearful, and violent. Users may go for long periods without food or sleep while under the effects of the drug. Stimulants are habit-forming. Some users switch between a stimulant and a depressant so they can sleep. A large dose of amphetamine or cocaine can cause the user's heart to stop beating.

Suggest some better ways to lose weight other than taking amphetamine drugs.

Hallucination
A distortion of the senses caused by mental disease or drugs

Psychedelic drug
Another term for hallucinogen

What Are Hallucinogens?

Hallucinogens confuse the central nervous system. They change the way the brain processes information from the senses. A person who uses a hallucinogen senses the world in a distorted way. These distortions of the senses are called **hallucinations**. The user might think that he or she hears voices. The user might become frightened of something that does not exist. Or the user might think that people are plotting against him or her.

Hallucinogens are also called **psychedelic drugs**. They are illegal and have no current accepted medical uses. Two hallucinogens, mescaline and psilocybin, grow naturally. LSD, phencyclidine (PCP), and MDMA (Ecstasy) are synthetic. A great danger of hallucinogens is that the user will hurt himself or herself or someone else. Users of PCP often become violent. A large dose of PCP can cause convulsions, coma, and death. Some hallucinogens cause flashbacks. The user experiences a hallucination. But it occurs months or years after the drug is taken.

LESSON 5 REVIEW Write the answers to these questions on a separate sheet of paper. Use complete sentences.

1) List three reasons why narcotics are dangerous.
2) Name two depressants and their effect on the body.
3) Name two stimulants and their effect on the body.
4) What kinds of drugs can cause hallucinations?
5) What is a synthetic drug?

Lesson 6

Other Dangerous Drugs

Inhalant
A substance that is breathed

In Lesson 5, you learned about four groups of drugs. Narcotics, depressants, and stimulants can be helpful when used for proper medical purposes. They are dangerous, however, when abused or used illegally. Hallucinogens are always dangerous and always illegal. This lesson describes the properties and effects of some other dangerous substances.

What Are Inhalants?

Breathable substances are called **inhalants**. You breathe, or inhale, the substance, which enters your body through your lungs. Some people use inhalants that contain medicine for asthma, severe allergies, or pain.

Some substances that are inhaled were never intended as medicines. These dangerous inhalants include household products such as glue, cleaners, paint removers, correction fluid, and gasoline. The labels on these products have warnings telling people not to breathe them.

People who ignore the warnings and breathe the products are abusing the products. In many places, this misuse of these products is illegal. People use these dangerous substances to get a hallucinogenic effect. Some people may simply be "experimenting."

Dangerous and illegal inhalants include paint thinner, polish remover, glue, and spray paint.

Using illegal inhalants is extremely dangerous. They depress the central nervous system and can cause slurred speech, poor judgment, confusion, dizziness, and headaches. Some illegal inhalants make people violent. Others can cause permanent brain damage. None of them are safe to use.

Anabolic steroid
A synthetic drug that resembles the hormone testosterone

Marijuana
An illegal drug from the hemp plant that produces intoxication

Sterile
Unable to have children

Steroid
A chemical that occurs naturally in the body or is made in a laboratory

Testosterone
The male hormone that produces male characteristics, such as facial hair and a deep voice

Why do young people use inhalants more often than other illegal drugs?

What Are Anabolic Steroids?

Steroids are chemicals, such as hormones, that occur naturally in the body or are made in a laboratory. Some people's bodies do not make enough of these needed chemicals. Then, a doctor can prescribe medicine to make up for the lack of the chemical.

One group of steroids is called **anabolic steroids**. These mimic the effects of the male hormone **testosterone**. This hormone produces characteristics, such as facial hair and a deep voice. People soon realized that taking anabolic steroids also stimulates the body to produce muscle tissue. Some athletes began to take these drugs to try to improve their performance. Young boys sometimes took the steroids because they wanted to appear older and stronger.

Using anabolic steroids is dangerous and is illegal without a prescription. Starting in the 1960s, athletes have been tested to see if they use steroids. If they do, they are disqualified from competition.

Anabolic steroids have many side effects. The liver, heart, and reproductive systems can all be permanently damaged. Steroids can cause cancer, heart attacks, and strokes. Both males and females can become **sterile**, or unable to have children. Teenagers who use steroids may stop growing and never reach their normal adult height. Steroids also have mental side effects. They can cause anxiety, depression, and violent behavior known as "roid rages."

What Is Marijuana?

Another illegal drug is **marijuana**. This drug comes from a hemp plant called Cannabis sativa. The leaves and buds of the plant have an intoxicating effect. To get this effect, some people smoke, eat, or make tea with this drug.

In the 1950s, some people believed that marijuana was safe to use. They were wrong. Marijuana depresses the central nervous system. It interferes with memory, concentration, and the ability to drive a car or do schoolwork. Marijuana smokers inhale deeply and hold the smoke in their lungs. This means that using marijuana can cause lung disease and cancer.

Designer drug
An illegal manufactured drug that is almost the same as a legal drug

Look-alike drug
An illegal manufactured drug that imitates the effect of other drugs

What Are Designer Drugs?

Criminals who make and sell illegal drugs try to avoid the law. One way they try is to create **designer drugs**. These are made from chemicals like those in a legal drug or medicine. The designer drugs, however, are different from legal drugs. They can be many times stronger and can imitate the effect of narcotics. Someone has created a designer drug called "China White." This drug has the same effect as heroin.

At one time, designer drugs were legal because the law had not described them as illegal. In 1986, a law called the Anti-Abuse Drug Act was passed. It made all designer drugs illegal. Designer drugs are dangerous because they often contain strong, poisonous chemicals.

What Are Look-Alike Drugs?

Another group of drugs made from legal chemicals are the **look-alike drugs**. They may imitate the effects of legal prescription drugs. However, look-alike drugs are illegally produced. A user cannot be sure what substances are contained in the drug. For example, someone may think he or she is buying an amphetamine, but the drug is actually LSD.

Look-alike drugs often contain dangerously large amounts of caffeine. They can cause anxiety, restlessness, weakness, headaches, and rapid heartbeat. These drugs are especially dangerous. The user cannot know how strong they are or what their side effects may be. The risk of taking an overdose is very high.

Health Tip

Use your refusal skills. Don't go to places where there will be drugs.

How can you use positive peer pressure to keep your friends from using drugs?

LESSON 6 REVIEW Write the answers to these questions on a separate sheet of paper. Use complete sentences.

1) What are three products that are misused as inhalants?
2) Why have some athletes used anabolic steroids?
3) What are the health risks of using anabolic steroids?
4) How does smoking marijuana affect the body?
5) Why are designer and look-alike drugs particularly dangerous?

Chapter Summary

- A drug changes the way the mind or body works. All medicines are drugs. Not all drugs are medicines.

- Prescription medicines require a written note from a doctor. Over-the-counter drugs do not need a doctor's prescription.

- The effects of medicines include relieving pain, preventing or curing disease, and helping the brain to function normally.

- Medicines affect people in different ways. Body size, weight, and other factors may change the effects of a medicine.

- Medicines may have side effects. Follow directions carefully when taking medicines.

- Do not take someone else's medicine. Do not take prescription medicine unless it was prescribed for your current illness.

- Tobacco is a stimulant. The nicotine in tobacco is addictive. The use of tobacco can result in life-threatening diseases involving the lungs, heart, and other organs.

- People who breathe secondhand tobacco smoke can have the same health problems that smokers have.

- Alcohol is a depressant. Misusing alcohol can result in the disease of alcoholism and cause permanent health and family problems.

- It is always dangerous to drive a car after drinking alcohol.

- Narcotics, depressants, and stimulants can all be addictive drugs. Illegal drugs in these groups are dangerous. They can cause permanent health damage and sometimes death.

- Hallucinogens change the way the brain processes information. All hallucinogens are illegal and dangerous.

- Inhalants, anabolic steroids, marijuana, designer drugs, and look-alike drugs are dangerous and addictive.

Chapter 9 Review

Comprehension: Identifying Facts

On a separate sheet of paper, write the correct word from the Word Bank to complete each sentence.

WORD BANK		
alcoholism	nicotine	stimulant
depressant	pain	tolerance
hallucinogenics	pregnant	tremors
inhalants	prescription	
narcotic	steroids	

1) A _____ is a drug that speeds up the central nervous system.

2) All forms of tobacco contain the addictive drug _____.

3) Alcohol is a central nervous system _____.

4) A drug user builds up _____ to a drug and needs more and more to get the same effect.

5) _____ drugs require a written note from a doctor.

6) People who stop using addictive drugs experience _____.

7) Morphine, cocaine, and heroin are _____.

8) All _____ drugs are illegal.

9) Women who are _____ must be particularly careful about using drugs.

10) The disease of _____ has no known cure.

11) One beneficial use of drugs is to prevent _____.

12) Paint thinner and glue are dangerous _____.

13) Some athletes abuse _____ to try to perform better.

Comprehension: Understanding Main Ideas

Write the answers to these questions on a separate sheet of paper. Use complete sentences.

14) List four safety tips for taking medicines.

15) What are two ways to decide which over-the-counter medicine to buy?

16) List drugs that can interfere with a person's ability to drive a car.

17) Why are tobacco and alcohol legal if they can cause harm to a person's health?

18) Name two narcotic drugs and two stimulant drugs that are illegal.

Critical Thinking: Write Your Opinion

19) Suppose you have a bad cold. Should you take medicine or just wait for the cold to go away? Explain your answer.

20) Why do you think alcoholism or any drug dependence could be considered a family disease?

Test Taking Tip Read test questions carefully to identify those questions that require more than one answer.

Chapter 10

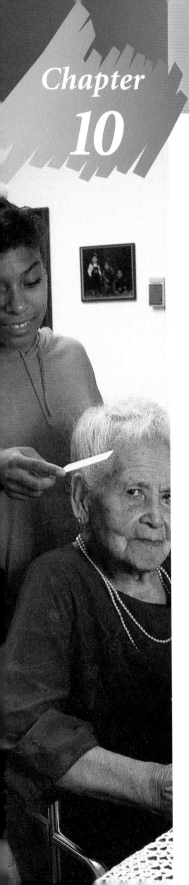

Drug Dependence—Problems and Solutions

As you grow older, you will be faced with many choices. The more you know about something, the better your choices will be. An important choice that you must make concerns the use of drugs. Both adults and teenagers are learning more about drugs. People now know how much damage drugs can do.

In this chapter, you will learn about drug dependence. Drug dependence causes problems with health. It causes problems with life. You will learn ways that drug users can get help. You will also learn about healthy alternatives to drug use.

Goals for Learning

▶ To explain the term *drug dependence*
▶ To describe how drug dependence affects a person's life
▶ To identify different ways that drug users can be helped
▶ To identify healthy activities to take the place of drug use

Lesson 1

The Problem of Drug Dependence

Drug abuse
The improper use of legal or illegal drugs

Drug dependence
The need for a drug that results from frequent use of that drug

Have you ever taken aspirin for a headache? When your pain stopped, did you want to continue taking aspirin? Probably not. People usually do not become dependent on aspirin. People can, however, become dependent on other kinds of drugs.

Dependence on alcohol and other drugs is a serious problem. It is a problem not only for individuals but for families and society.

Drug abuse includes the misuse or overuse of a medical drug. Drug abuse also includes use of *any* illegal drug.

What Is Drug Dependence?

Drug dependence is a need for a drug that results from frequent use of that drug. Sometimes, people become dependent on prescribed medicine. Most drug dependence, however, involves nicotine, alcohol, or other nonmedical drugs. Drug dependence often has three stages. In the first stage, the drug is used once in a while. In the second stage, the drug is used often. In the third stage, use of the drug takes over a person's life. Drug dependence is a disease. The only treatment is to stop using the drug.

A drug-dependent person may

- Use a drug often
- Use a drug in large amounts
- Use a drug although it causes serious problems

Drugs cause changes in the brain and body. A drug-dependent person builds tolerance. This means that the brain and body have gotten used to the drug. The user must take more of the drug to get the desired effect. As a user builds tolerance for a drug, he or she may take too much of the drug.

Healthy Subjects: Literature

"JUNIOR ADDICT"

The African-American writer Langston Hughes wrote many poems and stories about drug dependence. His poem "Junior Addict" describes an African-American boy. The boy is sad because of his drug use. His problem may be linked to the way his people have been treated in America. If only he can live to see "the sunrise" far away. The poet calls to the African sunrise to "come, quickly, come." The boy may find strength and hope if he can feel pride in his heritage.

Withdrawal
The body's physical reaction to the absence of a drug

What Are Some Signs of Drug Dependence?

If someone shows several of the following signs, they may have a problem with drugs.

- Major changes in appearance or behavior
- Loss of memory and concentration
- Lying, stealing, or borrowing money
- Reduced energy
- Loss of interest in favorite activities or hobbies
- Drop in performance at school or work
- Lateness and absences from school or work
- Trouble with the police
- Anger when discussing his or her drug use

What Is Withdrawal?

Another problem of drug dependence is **withdrawal**. Withdrawal is a physical reaction to the absence of a drug in the body. Withdrawal symptoms depend upon the kind of drug taken. If a person stops taking a stimulant, he or she might feel tired or sad. A person who stops taking a depressant might feel nervous. He or she might not be able to sleep. A person who is trying to stop using a drug should be under a doctor's care.

What Are the Effects of Drug Dependence?

Drug dependence causes many problems. Drug users often have problems with their family and friends. School and work suffer. The drug user's health becomes poor. He or she may have money problems. A user may even be arrested for having a drug or for crimes linked to drugs.

Health and Safety Problems

Many drugs damage the body. Different drugs can affect different parts of the body. Some drugs damage the major body systems. This damage can cause long-term health problems or even death.

Drug abuse increases the risk for injury and death. Alcohol and other drugs are related to more than 50 percent of traffic accidents in which someone dies. More than 69 percent of drownings involve use of alcohol or other drugs.

Family Problems

Drug abusers have problems with everyday life. Between 20 and 35 percent of all suicides are linked to drug abuse. Drug dependence damages families. To a drug abuser, the drug becomes the most important thing in his or her life. Drug-dependent parents may not be able to provide for their families. Families may not have enough money for food and a home.

A trained professional uses a combination of approaches that may best help each individual.

Drug-dependent people sometimes are violent. Almost one-half of all cases of wife abuse or husband abuse are linked to alcohol or other drugs. Likewise, drugs are linked to more than one-third of child abuse cases. The whole family suffers when one member is drug dependent.

Action for Health

LEARNING ABOUT THE COSTS OF DRUG USE

Find out how much drug abuse costs your community. Contact your county or state health department. Many cities and counties have mental health agencies and drug education agencies. Ask for information about drug abuse in your community.

Contact law enforcement agencies for information about crimes linked to drugs. Look in the city, county, and state government pages in the telephone book to find the phone numbers you need. You might call your local library for help.

Sexually transmitted disease
A disease spread by sexual contact

Problems for Society

Every year, drug abuse costs our country billions of dollars. Taxpayers pay more than $16 million a year to treat and support drug abusers. Businesses lose more than $98 million a year because of drug abuse by workers. The cost of crime linked to drugs is $25 million each year. These dollar costs are high. But they do not include the value of lives lost to drugs.

What Problems Are Linked to Drug Use by Teens?

For young people, drug use can cause problems that follow them through their whole life. Many teen drug users lose interest in school. They may not learn the basic thinking skills they need for life. Teen drug users often do not form normal relationships with people. Drug users have poor judgment. They often engage in high-risk behavior. As a result, the teen user may face an unwanted pregnancy. He or she may get a **sexually transmitted disease** or other diseases linked to drug use.

LESSON 1 REVIEW Write the answers to these questions on a separate sheet of paper. Use complete sentences.

1) What is drug abuse?
2) What is drug dependence?
3) What three tragedies are often linked to drug use?
4) What dangers do teen drug users face?
5) How does drug dependence affect family life?

Solutions to Drug Dependence

Detoxification
The removal of a drug from the body

Drug dependence is not a life sentence. People who are close to drug abusers can help them. Recovery from drug dependence is the first step.

How Can People Recover From Drug Dependence?

Recovery from drug dependence has three main steps. First, the drug-dependent person must admit that he or she has a problem with drugs. This seems easy, but it often is not. A drug user may deny that he or she is dependent on a drug. Friends and family members may have to confront the drug user.

The second step in recovery is **detoxification**. Detoxification is the removal of drugs from the body. To avoid serious problems, detoxification should take place in a hospital.

The third step in recovery is learning to live day-to-day without drugs. This is the longest process of recovery. The person in recovery learns skills to cope with life without drugs. He or she becomes emotionally and physically healthy. The process of recovery never stops. Drug dependence cannot truly be cured. Drug dependence can, however, be treated. The goal of recovery is never to use drugs again. Sometimes, people go back to using drugs. In this case, the process of recovery must begin again. With help, the recovering user still can reach the goal of a life without drugs.

Nutrition Tip

Take time to eat well when you're under stress.

How Can a Person Get Help?

There are several places to begin looking for help. If you or someone you are close to abuses drugs, ask for help.

Counseling

Most schools have counselors or a program to help students with problems. Students can discuss their concerns about drug use in their family or among their friends. Community mental health centers also can help. Mental health counselors can help people find treatment for their drug use. They can help people in recovery and help the families of recovering users.

Alateen
A support group for teens

Sponsor
A recovering alcoholic who helps a new AA member

Support group
A group of people with similar problems who help one another

Support Groups

A **support group** is a group of people with similar problems. The people in the group help and encourage one another. Alcoholics Anonymous (AA) was the first organization of support groups. The goal of AA members is to live a life without alcohol. Alcoholics must live day-to-day, choosing not to drink. In AA, each person has a **sponsor**. A sponsor is another recovering alcoholic. The sponsor usually has not used alcohol for several years. He or she helps the newly recovered alcoholic deal with the urge to drink.

AA has three other groups for family members of alcoholics. Al-Anon is a support group for adults, such as the husband or wife of an alcoholic. **Alateen** is for 12- to 18-year-olds. Alatot is for 6- to 12-year-olds. Members of these groups help one another deal with living with an alcoholic family member. Alateen also helps teens deal with alcoholic family members who don't live with them.

You can talk about drug use with a friend.

Other support groups help people deal with other kinds of drug dependencies. Two examples are Cocaine Anonymous and Narcotics Anonymous. The National Council on Alcoholism and Drug Dependence also can give information and help with problems linked to drugs.

Writing About Health

Think about some healthy choices you have made that kept you from drug use. Write about the choices. Did they have long-term physical health benefits? Did they benefit your mental health?

SMOKE ALARM

In the 1880s, a manufacturer used the first cigarette-making machine. Cigarette smoking spread. During and after World War II, advertising glamorized smoking.

Today, research shows that smoking contributes to heart and lung disease and cancer. Health warnings are now on advertisements and cigarette packages. Some states forbid smoking in public places. Airlines have banned smoking on U.S. flights.

Residential and Outpatient Treatment Centers

A residential treatment center is a place where recovering drug users live for a time. Some centers treat only teens. Others may treat all age groups. Some people get treatment from an outpatient center. In this case, recovering drug users live at home. They go to the center each day for treatment.

How Can a Person Stay Away From Drugs?

The main reason that teens use drugs is peer pressure. Sometimes it helps to practice saying "no" in a safe setting. Practice in a safe setting can help a person say "no" in a real setting. For example, a person may need to say "no" and leave a party where drugs are being used.

Refusing Drugs

If someone offers you drugs or pressures you to use drugs:

- Say "no," and walk away.
- Say "no," and change the subject.
- Say "no," and give honest reasons for not using drugs.
- Say "no," and keep repeating that you don't use drugs.

Getting Help

Getting help for problems can help a person avoid drug use. This is very important for someone from a home where drugs are used. Everyone is responsible for his or her own behavior. Family members are not to blame for the drug user's behavior.

Fitness Tip

Get regular exercise. You will be better able to cope with stress.

Drug Dependence—Problems and Solutions Chapter 10

DRUG ABUSE COUNSELOR

Drug abuse counselors work with people who are dependent on drugs. Some counselors are recovering drug users themselves. This helps the counselor understand the people they are helping. Counselors, together with other health professionals, develop treatment plans. Some counselors may work only with alcoholics. Others work with people addicted to other drugs. Some work only with children, teenagers, pregnant women, or other groups. Preparation may include a one-year certification program, two-year associate's degree, or four-year college degree.

What Are Some Better Choices Than Drugs?

Using drugs to deal with problems only makes the problems worse. Activities that help you feel better about yourself can help you. Here are some activities that can increase fitness and energy.

- Do something physically active. Try a dance class. Go mountain biking with a group. Join a swim team.

- Find a cause that is important to you and give your time to it. For example, volunteer at a hospital. Attend a beach cleanup day. Tutor a younger child in reading.

- Join a group of people with similar interests. For example, you could work on the school newspaper or join a school-sponsored club.

LESSON 2 REVIEW Write the answers to these questions on a separate sheet of paper. Use complete sentences.

1) What are the three steps in recovery from drug dependence?
2) Can drug dependence ever be cured?
3) How do support groups help people?
4) Where can a person go to get treatment for drug dependence?
5) How can you stay away from drugs?

> *Diseases have no eyes.*
> *They pick with a dizzy finger*
> *anyone, just anyone.*
> —Sandra Cisneros, *The House on Mango Street*

10) When a drug user stops taking a drug, he or she can have _____ symptoms.

11) A drug user going through recovery experiences _____, the removal of the drugs from the body.

12) Dangerous, illegal drugs that change the way the brain processes information are _____.

13) Drug users can live drug-free lives by going into _____ programs.

14) Treatment in a _____ involves drug users helping one another to stay drug free.

Comprehension: Understanding Main Ideas

Write the answers to these questions on a separate sheet of paper. Use complete sentences.

15) Why is it dangerous to ignore the directions on prescription medicine?

16) Make a list of legal drugs that are often misused.

17) What are some signs of drug dependence?

18) Describe the three stages of drug recovery.

Critical Thinking: Write Your Opinion

19) When is it safe to take drugs?

20) What kinds of refusal skills can help you avoid using dangerous drugs?

Unit 4 Review

Comprehension: Identifying Facts

On a separate sheet of paper, write the correct word or words from the Word Bank to complete each sentence.

WORD BANK		
abuse	inhalants	stimulant
depressant	narcotic	support group
detoxification	prescription	tolerance
hallucinogens	recovery	withdrawal
heroin	side effects	

1) Following the directions on medicines carefully will usually help prevent _____.

2) A doctor's written order for a drug is a _____.

3) Tobacco is a _____ that speeds up the central nervous system.

4) Alcohol is a _____ that slows down the central nervous system, making it dangerous to drive.

5) Morphine is a _____ drug used to relieve pain.

6) Misuse or overuse of a medical drug is called drug _____.

7) There are no legal uses for the narcotic drug _____.

8) Dangerous _____ include paint thinner and glue.

9) Drug users can build up _____ so that they need more and more of the drug.

Unit Summary

- Medicines are drugs used to treat, cure, or prevent disease.

- Medicines affect different people differently because of body size, weight, or other factors.

- All medicines can have side effects. Following directions on the label carefully will usually prevent problems.

- Tobacco is a stimulant drug that can cause life-threatening diseases. The nicotine in tobacco is addictive. Secondhand tobacco smoke is also dangerous to people's health.

- Alcohol is a depressant drug that can be addictive and is often misused. Alcohol abuse can cause serious diseases and contributes to many traffic fatalities.

- Most narcotics, depressants, and stimulants are legal drugs, but they can be misused or bought illegally.

- Hallucinogens are particularly dangerous drugs that change the way the mind processes information.

- Inhalants, anabolic steroids, marijuana, designer drugs, and look-alike drugs are dangerous and addictive.

- A person dependent on drugs must take more and more of the drug to get the same effect. Signs of drug dependence include changes in behavior, lying or stealing, and poor work habits.

- Drug users often engage in high-risk behavior. For example, injecting drugs with used needles can lead to diseases such as AIDS.

- People can recover from drug dependence. They must admit they have a problem, go through detoxification, and then learn to live without drugs.

- Support groups for drug users and their families can help people change to drug-free lives.

- The best way to avoid drug problems is never to use drugs. Avoiding people who use drugs and saying "no" to drugs are ways to stay drug free.

Deciding for Yourself

Using Drugs to Change Emotions

There are many drugs that can change how a person feels. Some of these are tranquilizers such as Valium and Librium. These drugs calm a person down. Other drugs are stimulants. They can help a depressed person feel more hopeful. Drugs that change emotions and mental states can be helpful when used correctly.

Drugs are one way to change emotions. Taking a drug may sound like an easy solution to someone who is suffering mental or emotional pain. But drugs may not be the best answer.

People can become dependent on taking drugs. They may come to believe they need the feelings and emotions the drugs provide. Then they take the drug more and more frequently. They may become addicted to the drug.

Taking drugs to change emotions can prevent someone from finding the cause of those emotions. If you felt anxious, you could take a drug. But the drug is only a temporary solution. When the drug wears off, the problems causing the anxiety come back. Rather than taking drugs, it is better to find out what is causing the depressed emotions and deal with the underlying problem.

There are other ways to change emotions. Positive action and positive thinking can help with depression. Physical activity can reduce stress. Talking with someone about your feelings and emotions can help you deal with them.

Questions

1) What kinds of drugs are used to change emotions?
2) Why might it be a bad idea to take drugs for emotional problems?
3) What should someone do before taking a drug to help with an emotional or mental problem?
4) Why do you think some people use drugs rather than other solutions for mental and emotional problems?

11) The second step of recovery from drug dependence is _____.

12) Members of a _____ help one another with problems.

13) A support group for teens is called _____.

Comprehension: Understanding Main Ideas

Write the answers to these questions on a separate sheet of paper. Use complete sentences.

14) On what kinds of drugs do people usually become dependent?

15) What are three signs of drug dependence?

16) Why is drug abuse particularly harmful to teens?

17) What is the first step of recovery, and why is it sometimes difficult?

18) If you live with a person, is it your fault that they are drug-dependent? Why or why not?

Critical Thinking: Write Your Opinion

19) Why do you think finding positive activities helps people stay away from drugs?

20) What could you do to help a drug-dependent friend or family member?

Test Taking Tip Make a short outline of the main ideas of the chapter using the paragraph headings that appear in the text.

Chapter 10 Review

Comprehension: Identifying Facts

On a separate sheet of paper, write the correct word or words from the Word Bank to complete each sentence.

WORD BANK	
Alateen	sexually transmitted disease
alcohol	support group
behavior	teen
brain	teen drug users
detoxification	tolerance
drug abuse	withdrawal
drug dependence	

1) Frequent use of a drug is called _____.

2) Misuse or overuse of a medical drug, or any use of an illegal drug, is called _____.

3) _____ and other drugs are related to more than 50 percent of traffic accidents in which someone dies.

4) The buildup of _____ means that a drug user must use more and more of the drug.

5) When a drug-dependent person stops a drug, he or she may go through _____.

6) A major change in appearance or _____ may signal drug dependence.

7) Many _____ lose interest in school.

8) Drugs often damage the _____.

9) Drug abusers may engage in high-risk behavior. Therefore, some may get a _____.

10) _____ drug users are more likely to use many drugs later in life.

222 Chapter 10 Drug Dependence—Problems and Solutions

Chapter Summary

- Drug dependence is a need for a drug resulting from frequent use of that drug. Most drug dependence involves nicotine, alcohol, or other nonmedical drug.

- Drug abuse includes the misuse or overuse of a medical drug. It also includes use of any illegal drug. Tolerance to a drug develops when the brain and body become used to the drug. Then the user must take more of the drug to get the desired effect.

- When the drug is stopped, withdrawal symptoms occur. Withdrawal is a physical reaction to the absence of a drug.

- Signs of drug dependence include changes in behavior or appearance, lying or stealing, loss of interest in old activities, and/or poor work habits.

- Drugs damage the body. Drugs can cause long-term health problems or death.

- Drugs play a part in many traffic accidents, drownings, and suicides. Many cases of abuse are linked to drug use.

- Drug users often engage in high-risk behavior. Unwanted pregnancy and getting sexually transmitted diseases are sometimes linked to drug use.

- People can recover from drug dependence. The first step is admitting that they have a problem with drug use.

- The second step in recovery is detoxification, which is removal of drugs from the body. The third and longest step is learning to live without drugs.

- There are several ways a drug-dependent person can get help. Mental health counselors can help people find treatment.

- Support groups can help drug-dependent people realize that they are not alone. Support groups for families of drug users help members deal with living with a drug-dependent person.

- The best way to avoid drug problems is never to use drugs. You can avoid people who use drugs. You can say "no" to an offer of drugs and mean it.

Unit 5

Preventing and Controlling Diseases and Disorders

*I*f you are rarely sick, you are fortunate. You may occasionally have a cold or a headache. If this is true for you, you may have made smart decisions about your health. Developing good health habits can prevent some diseases. It helps you recover more quickly when you do get sick.

Diseases are caused in different ways. In this unit, you will learn about the causes of different types of diseases. You will read about ways these diseases can be cured or prevented. You will also find out things you can do right now to decrease your chances of getting serious diseases.

▶ Chapter 11 Disease—Causes and Prevention

▶ Chapter 12 Preventing AIDS and Sexually Transmitted Diseases

▶ Chapter 13 Common Diseases

Ryan White, 1972–1990

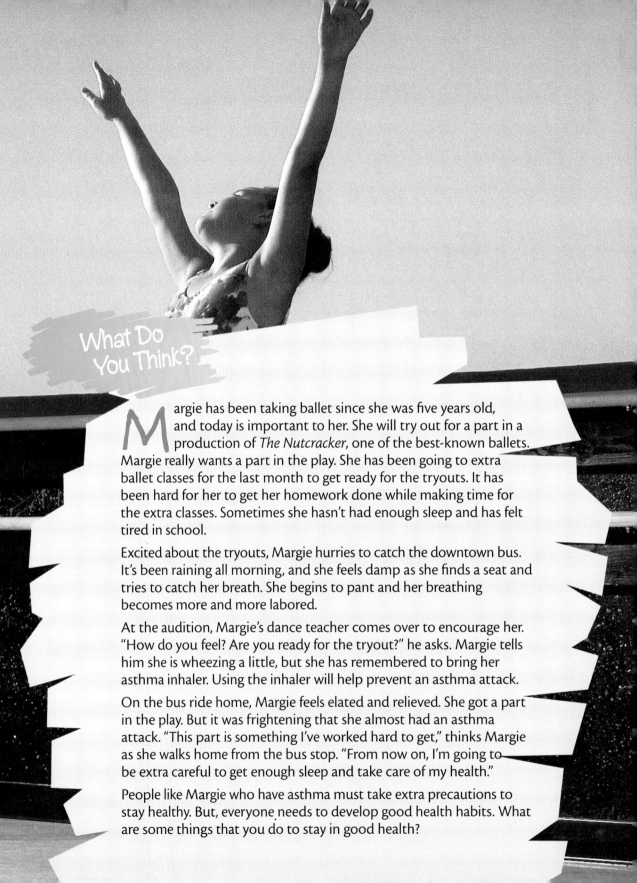

What Do You Think?

Margie has been taking ballet since she was five years old, and today is important to her. She will try out for a part in a production of *The Nutcracker*, one of the best-known ballets. Margie really wants a part in the play. She has been going to extra ballet classes for the last month to get ready for the tryouts. It has been hard for her to get her homework done while making time for the extra classes. Sometimes she hasn't had enough sleep and has felt tired in school.

Excited about the tryouts, Margie hurries to catch the downtown bus. It's been raining all morning, and she feels damp as she finds a seat and tries to catch her breath. She begins to pant and her breathing becomes more and more labored.

At the audition, Margie's dance teacher comes over to encourage her. "How do you feel? Are you ready for the tryout?" he asks. Margie tells him she is wheezing a little, but she has remembered to bring her asthma inhaler. Using the inhaler will help prevent an asthma attack.

On the bus ride home, Margie feels elated and relieved. She got a part in the play. But it was frightening that she almost had an asthma attack. "This part is something I've worked hard to get," thinks Margie as she walks home from the bus stop. "From now on, I'm going to be extra careful to get enough sleep and take care of my health."

People like Margie who have asthma must take extra precautions to stay healthy. But, everyone needs to develop good health habits. What are some things that you do to stay in good health?

Chapter 11

Disease—Causes and Prevention

*I*f you are healthy, you probably don't think much about your health. But when you are sick, health seems precious. And it is. Many of the things you do affect your health. Healthy habits can improve your life by guarding your health.

In this chapter, you will learn about the causes of some common diseases. You will learn how your body fights disease. You will also learn how to prevent many diseases. Forming good habits at a young age will affect your health for years to come.

Goals for Learning

▶ To explain the difference between inherited and acquired diseases

▶ To describe how pathogens cause disease

▶ To explain how the inflammatory and immune responses protect the body

▶ To explain how immunizations prevent diseases

Lesson 1: Causes of Disease

Think back to the last time you were sick. Perhaps you had a cold or the flu. You might have diabetes or asthma, **diseases** that do not go away. Disease is the interruption or disorder of normal body operations. A person can either inherit or acquire diseases.

What Are Inherited Diseases?

Inherited diseases run in families. They are caused by **genes** that are passed from parent to child. A gene is like a recipe for making part of the body. If the recipe is wrong, the body will not behave normally. You might have heard of some inherited diseases. **Muscular dystrophy** is an inherited disease. In a person with muscular dystrophy, the muscles do not develop normally.

What Are Acquired Diseases?

An **acquired disease** is not caused by genes. Acquired diseases are caused by contact with germs, human behaviors, or the environment. For example, lung cancer is an acquired disease. Most lung cancer is caused by smoking. In some cases, the environment can affect health and cause acquired diseases. Some diseases, such as heart disease, are partly inherited and partly acquired. The tendency for heart disease is inherited. However, a person's lifestyle can prevent the disease from developing.

Germs that cause acquired diseases are called **pathogens**. Think back to that cold you had. You might have said, "I caught a cold." In a sense, this is true. You probably picked up cold germs on your hands. You might have breathed them in from the air. Pathogens are everywhere. However, you can lower your chances of getting a cold by using healthy habits.

Acquired disease
A disease caused by infection or human behavior

Disease
A disorder of normal body function

Gene
Parts of a cell that are passed from parent to child

Inherited disease
A disease passed through genes

Muscular dystrophy
An inherited disease in which the muscles do not develop normally

Pathogen
A disease-causing germ

Writing About Health

Think about your personal health practices. List the things you could do to prevent getting an infectious disease such as a cold. Then list the things you could do when you have a cold to avoid spreading it to others.

Infection
A sickness caused by a pathogen in the body

Infectious disease
A disease caused by a pathogen

Mucous membrane
The thin, moist tissue that lines body openings

Mucus
The sticky fluid produced by mucous membranes

Multiply
Increase in number

How Does the Body Stop Pathogens?

Your body has several roadblocks for pathogens. **Mucous membranes** are the moist surfaces that line your mouth, nose, and other body openings. Mucous membranes make a sticky liquid, **mucus**. Mucus traps pathogens before they enter the body. Hairs in the nose trap pathogens. Acid in the stomach kills many pathogens. Coughing, sneezing, and even vomiting remove pathogens from the body.

Some pathogens do, however, enter the body. A cut in the skin is a way in. Pathogens can also enter your body in food. If pathogens do enter your body, they may **multiply**. This process is called **infection**. Diseases caused by pathogens are called **infectious diseases**. Some infectious diseases, such as a cold, are not serious. Others, such as AIDS, are deadly. Figure 11.1 shows the stages of an infectious disease.

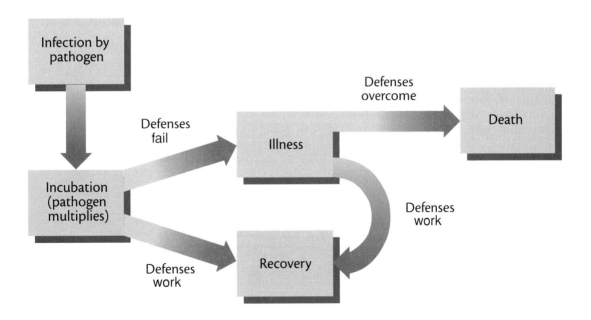

Figure 11.1. Stages of infectious disease

HEALTH SERVICE COORDINATOR

Health service coordinators work in doctors' offices, clinics, and hospitals. They keep information about patients. They give information to nurses, doctors, and other health care workers. They order special diets, drugs, equipment, supplies, laboratory tests, and X-ray exams. Health service coordinators may have different job titles depending on their specific job. Two to four years of training are required. Health service coordinators also must know basic medical terms and nursing and testing procedures.

How Can I Avoid Getting an Infectious Disease?

Many pathogens are passed from person to person on the hands. Wash your hands often. Always wash your hands before you eat or prepare food. Wash your hands after you use the bathroom or take out the garbage. Pathogens live by the millions on doorknobs, telephone receivers, and stair rails. In public places such as airports, there are more pathogens than there are in your home. Assume that you have pathogens on your hands unless you have just washed them. Avoid rubbing your eyes or touching your lips with unwashed hands.

Some pathogens live in uncooked meat and eggs. When you prepare food, cook meats until they are well done. Do not eat food containing raw eggs. Clean kitchen surfaces and utensils with detergent and hot water before you use them again. Once you have learned how infectious diseases are spread, it becomes easy to avoid many of them.

Fitness Tip

Exercise will increase your resistance to infectious diseases.

Health Tip

Wash your hands often to prevent some infectious diseases.

LESSON 1 REVIEW Write the answers to these questions on a separate sheet of paper. Use complete sentences.

1) What is an inherited disease?
2) What is an acquired disease?
3) What do pathogens do when they enter your body?
4) Describe some ways the body blocks pathogens.
5) How can washing your hands reduce your chances of getting an infectious disease?

Lesson 2
The Body's Protection From Disease

Immune system
A system of organs, tissues, and cells that fight infection

Inflammatory response
The body's first response to a pathogen

Some pathogens get through the body's protection from infection. A healthy body is ready, however, to fight pathogens in several ways.

What Happens When Pathogens Invade?

When you are infected by pathogens, the body tries to fight the infection. If you have ever cut your finger, you are familiar with some of the ways the body fights. The area around the cut swells. Blood vessels become larger. The body sends special blood cells to the injured area. These blood cells kill pathogens. The injured area becomes hotter than the rest of the body. The heat slows the multiplication of the pathogens. These events are called the **inflammatory response**. They are the first things that happen when pathogens enter.

The inflammatory response may not stop the pathogens. They may still multiply and start an infection. The body, however, does not give up. A special system of organs, tissues, and cells begins to fight pathogens. This system is called the **immune system**.

Healthy Subjects
Social Studies

THE BUBONIC PLAGUE

An *epidemic* is a widespread disease. Some of the world's worst epidemics were of bubonic plague. Bubonic plague is caused by a pathogen carried by fleas. The fleas move from place to place on the bodies of rats. Bubonic plague causes fever, pain, and open sores. Before modern medicine, the plague was usually deadly. The first recorded plague was in 435 B.C. The worst known outbreak occurred in Europe from 1347 to 1351. Known as the Black Death, it killed millions of people. In the early 1900s, over ten million people in India died of the plague. Today, there are medicines to fight this disease. Over the years, a few cases of plague have appeared in the United States. When this happens, the U.S. Public Health Service isolates victims and cleans up infected areas. This stops the spread of the disease.

Immune
Resistant to infection

Immunization
Means of making a body immune from a disease

Vaccination
An injection of dead or weakened viruses to make the body immune to the virus

How Does the Immune System Work?

The immune system is the body's final line of defense against infection. When a foreign substance enters the bloodstream, the blood make antibodies. Antibodies are proteins that kill specific pathogens. If the correct antibodies are in the blood when an infection occurs, they begin to fight it. They may knock out the infection early enough so that you don't become sick.

Antibodies form in the blood in a number of ways. If you do become sick, antibodies to a pathogen develop during the course of the illness. Then the antibodies help to fight the illness so that recovery is speeded up.

Some antibodies remain in the body after a disease is over. These antibodies give you immunity so you won't get the disease again. You are **immune**.

How Does a Vaccination Work?

Antibodies may also be in the blood as a result of **vaccination** or **immunization**. A vaccination is an injection, a shot, or oral medication of dead or weakened viruses into your body. Enough virus is injected to make you immune to but not sick from the disease. For example, when you get a measles vaccination, your body makes antibodies to the measles virus. Then you are immune to measles.

TONSILS

Tonsils are pieces of tissue that lie on either side of the back of the throat. Sometimes tonsils become infected. This disease is called *tonsillitis*. Tonsillitis can cause a sore throat, fever, and difficulty swallowing. For many years, doctors simply removed diseased tonsils.

In recent years, however, researchers have learned that tonsils are useful. Tonsils are part of the body's immune system. They trap pathogens entering the body. Also, the tonsils produce chemicals that help fight infections. Today the common treatment for tonsillitis is medicine. Tonsils are sometimes still removed. But that happens only when the infection doesn't go away with medical treatment.

Why might a person in poor general health get several different diseases?

You already have antibodies when you are born. You are born immune to some diseases. For example, humans cannot get certain diseases that animals get. Humans are naturally immune to such diseases.

Some immunities are acquired naturally. Babies are protected from certain diseases because of antibodies in their mother's milk. This immunity works in babies until their own immune system takes over. Without the immune system, people would have many more diseases and infections. Your immune system is an important defense against disease.

The table on the next page shows some of the vaccines that are given to people.

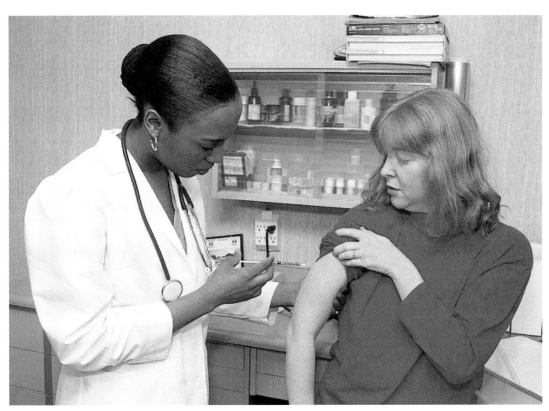

Vaccinations help the body build immunity to some diseases.

Action for Health

KEEPING TRACK OF YOUR VACCINATIONS

Ask your family to help you make a list of all your vaccinations and the dates when you had them. This chart shows some usual types of vaccinations and the ages when they are recommended. If your vaccinations are not up-to-date, see your doctor.

Vaccine	Recommended Age
Diphtheria, whooping cough, tetanus	2, 3, 6, and 15 months; 4 to 6 years
Oral polio vaccine	15 months; 4 to 6 years
Measles	15 months; 4 to 6 years
Mumps	15 months; 4 to 6 years
German measles	15 months; 4 to 6 years
Tetanus, diphtheria	14 to 16 years; every 10 years thereafter
Chicken pox	12 to 18 months; adults who have never had the disease
Hepatitis B	three injection series; birth through adults; usually by age 11

LESSON 2 REVIEW Write the answers to these questions on a separate sheet of paper. Use complete sentences.

1) What happens during the inflammatory response?
2) How does the immune system work?
3) Where are antibodies made?
4) How do we build antibodies to pathogens?
5) What is a vaccination, and what does it do in the body?

Chapter Summary

- Diseases can be either acquired or inherited.

- A person gets an inherited disease through genes.

- Infection, human behaviors, or environmental conditions cause acquired diseases. An acquired disease can be picked up from contact with germs.

- Some diseases are partly inherited and partly acquired.

- Pathogens are germs that cause acquired diseases. Pathogens can enter the body through a cut in the skin or in food. Once in the body, they multiply and cause infection.

- The body has several protections from infections. The skin, mucous membranes, stomach acids, coughing, sneezing, and vomiting help prevent infection.

- The inflammatory response stops infections by some pathogens.

- Good health habits can help avoid infection. Washing hands, cleaning kitchen surfaces, and cooking food well help to prevent infectious diseases.

- The immune system fights foreign substances in the bloodstream by producing antibodies.

- Antibodies get in the blood in a number of ways. Earlier infections or vaccinations build antibodies in the blood to fight infections. You are born with certain antibodies. A baby gets them through its mother's milk.

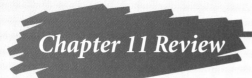

Chapter 11 Review

Comprehension: Identifying Facts

On a separate sheet of paper, write the correct word or words from the Word Bank to complete each sentence.

WORD BANK	
acquired disease	infectious disease
antibodies	inflammatory response
blood	inherited diseases
disease	mucous membranes
gene	mucus
immune system	pathogens
immunization	vaccination
infection	

1) When you have a _____, the body does not work normally.
2) _____ run in families.
3) A _____ causes an inherited disease.
4) Germs that cause acquired diseases are called _____.
5) The moist surfaces that line the mouth and other body openings are _____.
6) _____ is a sticky fluid that traps pathogens.
7) Multiplication of pathogens in the body is called _____.
8) An _____ is a disease caused by pathogens.
9) During the _____, blood vessels get larger and the area gets hot.

10) The _____ is a system of special organs, tissues, and cells that fights disease.

11) Lung cancer is an _____.

12) Heart disease is an _____ and an _____.

13) _____ makes antibodies.

14) _____ fight diseases by killing specific pathogens.

15) _____ is a way to prevent people from getting infectious diseases.

16) _____ uses dead or weakened viruses.

Comprehension: Understanding Main Ideas

Write the answers to these questions on a separate sheet of paper. Use complete sentences.

17) Do you have more control over getting an inherited disease or an acquired disease?

18) How do people get antibodies from immunizations?

Critical Thinking: Write Your Opinion

19) What might happen if the body made a mistake and attacked its own cells?

20) Do you think immunizations should be required of people living in society? Why or why not?

Test Taking Tip Be sure you understand what the test question is asking. Reread it if necessary.

Chapter 12

Preventing AIDS and Sexually Transmitted Diseases

Some diseases are spread by contact during sexual activity. A pregnant woman with a sexual disease can give it to her baby. Some of these diseases can be cured, but others cannot. All sexually transmitted diseases can cause serious health problems.

In this chapter, you will learn about what causes some sexually transmitted diseases. You will read about the life-threatening disease AIDS. The lessons in the chapter will describe the symptoms and treatments for these diseases. You will also learn how sexually transmitted diseases can be prevented.

Goals for Learning

▶ To describe the causes and symptoms of AIDS

▶ To identify the ways that AIDS is transmitted

▶ To describe the symptoms and treatment of four other sexually transmitted diseases

▶ To learn how sexually transmitted diseases can be prevented

Lesson 1

AIDS

AIDS
Acquired immunodeficiency syndrome, a disorder of the immune system

HIV
Human immunodeficiency virus, the virus that causes AIDS

AIDS is a serious disease of the immune system. The letters in AIDS stand for acquired immunodeficiency syndrome. The disease is acquired because one person can acquire it, or catch it, from someone else.

The immune system is what the body uses to fight off infections and diseases. If a person has AIDS, the immune system is damaged or weakened. So, the person is at risk of getting many serious diseases. Nearly all people with AIDS will die from one of these diseases. Right now, there is no cure for AIDS.

What Causes AIDS?

AIDS is caused by a virus called the human immunodeficiency virus, or **HIV** for short. HIV invades the body through the bloodstream. It infects body cells and cripples the human immune system. People who have HIV in their blood do not always have AIDS. But most people with the virus will eventually get AIDS.

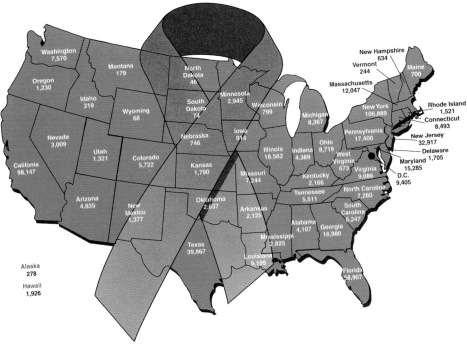

Figure 12.1 Number of reported cases of AIDS. Numbers show the total cases through December of 1996.

Preventing AIDS and Sexually Transmitted Diseases Chapter 12 **243**

THE SPREAD OF AIDS

In 1984, about 3,450 people died from AIDS. The number of deaths increased each year. In 1996, there were 50,140 deaths. Then, new drugs and prevention programs began to work. In 1997, the number of deaths was down to 38,780.

Many scientists are working to find treatments and cures for AIDS. Because of research and greater health precautions, people who now have AIDS are living longer and have fewer painful symptoms. But AIDS is still a disease with no cure. Until scientists find a cure, the only protection against AIDS is not to get the disease.

HIV negative
Not having HIV in the blood

HIV positive
Having HIV in the blood

Kaposi's sarcoma
A rare type of cancer affecting the skin or internal organs

How Is AIDS detected?

If a doctor suspects that someone has AIDS, the doctor orders a blood test. The test can find out if HIV is in the person's blood. A person with HIV is **HIV positive**. This means that the test was positive. It showed that the blood has HIV. If someone does not have the virus, the person is **HIV negative**.

What Are the Symptoms of AIDS?

AIDS is an illness that can have a long incubation period. This means a person with HIV may not show signs of AIDS for six to ten years. During all that time, the person appears healthy. But the disease is weakening the person's immune system.

Eventually, the symptoms of AIDS will appear. These symptoms include swollen lymph nodes, weight loss, skin rashes, diarrhea, fever, and night sweats. As the disease gets worse, people with AIDS can develop memory loss, personality changes, and paralysis.

Because HIV weakens the immune system, people with AIDS can have serious diseases. For example, they may get a lung infection called pneumonia. They may get a rare type of cancer called **Kaposi's sarcoma**. It affects the skin or internal organs.

Communicable
Able to be passed from one person to another

Transfusion
The transfer of blood from one person to another

How Do People Get AIDS?

Like all illnesses caused by a virus, AIDS is a **communicable** disease. This means that one person can get the disease from another. HIV, which causes AIDS, is in the blood and other body fluids of the sick person. These fluids can carry the disease to another person. There are several ways that someone can get HIV.

Sexual Contact

AIDS is spread through sexual activity. During sexual activity, body fluids, such as semen and vaginal secretions, are passed from one person to the other. If one person has HIV, the partner can get it during sex.

Infected Blood and Needles

Any needle used for injecting a drug or medicine gets blood on it. If a person with HIV is injected, then afterwards, that needle can carry HIV. Sometimes drug addicts share needles. In this way, HIV is passed from one person to another.

Sometimes a sick or an injured person needs more blood. The person receives a **transfusion** in which blood is transferred from one person to another. Before it was known how HIV spread, some infected people donated blood for use in transfusions. The infected blood carried HIV to new victims. A blood test for AIDS was developed in 1985. Now the American blood supply is almost always safe.

Healthy Subjects
Social Studies

A hospice is a nontraditional setting for health care. People dying with diseases such as AIDS can choose to live in a hospice to get the care they need. Some patients in hospices can no longer be helped in a hospital.

A hospice has a homelike setting. Family, friends, and volunteers assist those with medical training in a hospice. The volunteers may feed or dress patients. They may read and talk to them.

Mother to Child

A woman can pass HIV to her baby before, during, or after birth. The virus can also be passed to the baby while the baby drinks the mother's milk. Pregnant women can be tested for HIV. There are now ways to prevent the transfer of the disease to the unborn baby.

What Are Some Myths About AIDS?

AIDS is a serious disease that cannot be cured. Because of this, some people are afraid to be around a person who has AIDS. It is important to understand that AIDS is not acquired by breathing the air people with AIDS breathe. Touching, holding, hugging, or shaking hands with a person with AIDS will not spread the disease. Sharing dishes or swimming in a pool with an AIDS victim will not spread the disease either. It is safe for children with AIDS to go to school with other children. The other children are not at risk of catching the disease.

Children with AIDS are the innocent victims of this deadly disease.

LESSON 1 REVIEW Write the answers to these questions on a separate sheet of paper. Use complete sentences.

1) What is the difference between HIV and AIDS?
2) What are the symptoms of AIDS?
3) How is AIDS passed from one person to another?
4) How can AIDS be prevented?
5) Is it safe to be around a person with AIDS? Explain your answer.

If a student in your school has AIDS, do the other students need to be tested? Why or why not?

Sexually Transmitted Diseases

Chronic
Lasting

Genital herpes
A sexually transmitted chronic infection

Sexually transmitted disease
A disease passed from one person to another through sexual contact

Any disease that can be spread through sexual activity is a **sexually transmitted disease**. One example is AIDS. Other diseases pass from one person to another through sexual contact. Most sexually transmitted diseases can be cured if they are discovered early.

Which Sexually Transmitted Diseases Have No Cure?

Two sexually transmitted diseases have no cure. There are no vaccines to prevent these diseases. Once a person has the disease, there is no way to eliminate it from the body.

AIDS

You read about the AIDS disease in Lesson 1. AIDS has no cure. Almost all people who get HIV, which causes AIDS, will die from the disease. AIDS passes from one person to another through sexual contact or through needles used by infected people.

Genital Herpes

A less serious incurable disease is **genital herpes**. This is a **chronic**, or lasting, infection. The main symptom of genital herpes is clusters of painful, small blisters in the genital area. The blisters break, heal, and come back.

Genital herpes is spread by contact with the broken blisters. Avoiding sexual contact with an infected person can prevent someone from getting this disease.

Why is it difficult to prevent the spread of sexually transmitted diseases?

Genital herpes can lead to other health problems. Women may get repeated infections involving the cervix. This can lead to cancer of the cervix. A pregnant woman can pass genital herpes to her baby. During birth, the baby passes through the birth canal. The baby then becomes infected with the disease. Genital herpes can cause brain damage in babies.

Genital herpes has no known cure. A pill can speed up healing but does not get rid of the infection. Over time, the infections tend to be less severe. But genital herpes never goes away.

Action for Health

PRACTICE SAYING "NO"

Some teenagers experience peer pressure to have sexual activity. But it is a smart decision to decide not to have sex. If you decide to abstain from sex, you may need to tell others of your decision. You will need to express yourself in ways that are very clear to others so that they understand your wishes. Make a list of different ways you can communicate "no" to someone who thinks you should have sexual relations.

Gonorrhea
A sexually transmitted disease that often has no symptoms

Penicillin
An antibiotic used to treat diseases such as gonorrhea

What are Other Sexually Transmitted Diseases?

Other sexually transmitted diseases can be cured if they are discovered early. But if these diseases are not treated, they can lead to other serious health problems.

Gonorrhea

The most commonly reported infectious disease in the United States is **gonorrhea**. At least two million cases occur each year. Gonorrhea is sexually transmitted. It can also be passed to babies of infected women as the babies pass through the birth canal. Gonorrhea can cause an eye infection in babies that may lead to blindness.

In males, the symptoms are a white discharge from the urethra and burning while urinating. Females may have a vaginal discharge and some swelling and redness in the genital area.

Native Americans often contracted sexually transmitted diseases from Europeans. Why were the diseases particularly harmful to Native Americans?

Gonorrhea is dangerous because many infected people have no symptoms. Most females have no symptoms. From 20 to 40 percent of males have no symptoms either. If gonorrhea is not treated, it can cause sterility. Sterility is the inability to have children.

Gonorrhea can be treated successfully with the antibiotic **penicillin**. Penicillin destroys the bacteria that cause the disease. If a person thinks he or she may have gonorrhea, many health clinics can diagnose and treat the disease.

Chlamydia
A sexually transmitted disease that often has no symptoms

Syphilis
A sexually transmitted disease that has three stages

Health Tip

If you avoid sexual contact, you have almost no chance of getting a sexually transmitted disease.

Chlamydia

The symptoms of **chlamydia** are like those of gonorrhea. There may be a discharge from the urethra in males or from the vagina in females. Usually, however, females have no symptoms. Like gonorrhea, this disease can pass from a mother to her baby during birth. Chlamydia can cause an eye infection in babies that leads to blindness. Chlamydia can be treated with an antibiotic.

Syphilis

If **syphilis** is not treated, it passes through three stages. In the first stage, there is a small, painless, hard sore with a small amount of yellow discharge. The sore appears on the penis, anus, or rectum in men; on the cervix and genital areas in women. In the second stage, there is a rash and sometimes headache, loss of appetite, and other symptoms. During the third stage, the symptoms disappear for many years. But the disease is still there. It can damage the heart and brain.

Syphilis is passed from one person to another through sexual contact. It can pass from a woman to her baby before the baby is born. Syphilis is easily diagnosed by a blood test. A medical professional can also look through a microscope at a sample taken from a sore. Syphilis can be treated with antibiotics.

How Are Sexually Transmitted Diseases Prevented?

All newborns are given eye medicine after birth to prevent eye infections that might be caused by sexually transmitted diseases.

Most states require people to have blood tests before they can get a marriage license. The test identifies diseases such as syphilis. Then the diseases can be treated with antibiotics such as penicillin.

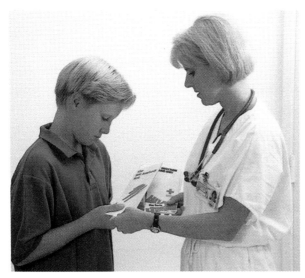

Talk to a medical professional immediately if you think you have a sexually transmitted disease.

Careers

LABORATORY ASSISTANT

Do you like to look through microscopes at tiny organisms? Then you might want to be a laboratory assistant. Laboratory assistants work with medical technicians. They look at samples of human fluids and tissues for organisms that cause disease. They report all situations that do not appear normal so that doctors can treat diseases. Laboratory assistants also help to keep medical equipment working properly. They must also protect themselves from disease. They wear disposable gloves and other protective clothing. You need one or two years of college to become a laboratory assistant.

Many health clinics emphasize the diagnosis and treatment of sexually transmitted diseases. The people who work at these clinics know that the possibility of having such a disease can be embarrassing. They protect the identity of their patients.

LESSON 2 REVIEW Write the answers to these questions on a separate sheet of paper. Use complete sentences.

1) Which sexually transmitted diseases have no cure?

2) Which sexually transmitted diseases can be passed from a woman to her baby?

3) Why is it important to diagnose sexually transmitted diseases quickly?

4) Which diseases mentioned in this lesson often have no symptoms?

5) How can the spread of sexually transmitted disease be stopped?

Writing About Health

What would you say to a friend who is thinking about starting a sexual relationship? Write how you could warn your friend about the dangers of this activity.

Chapter Summary

- AIDS (acquired immunodeficiency syndrome) is a communicable and life-threatening disease. It is caused by HIV (human immunodeficiency virus).

- HIV, which causes AIDS, is spread through certain body fluids.

- A person can contract AIDS during sexual activity or by using an infected needle.

- A woman with AIDS can pass the disease to her baby.

- People previously got AIDS through blood transfusions. Now the U.S. blood supply is safer because blood can be tested for HIV.

- The first symptoms of AIDS are minor. As the disease progresses, the person can get serious diseases such as pneumonia and Kaposi's sarcoma.

- A person with HIV may not show any symptoms for six to ten years.

- There is no cure for AIDS. It can be prevented by avoiding contact with body fluids that may be infected with HIV.

- A sexually transmitted disease is spread through sexual contact.

- Two sexually transmitted diseases that are not curable are AIDS and genital herpes.

- The symptoms of genital herpes are painful blisters in the genital area.

- Gonorrhea often has no symptoms. It can be treated with the antibiotic penicillin. If not treated, it can cause sterility in adults or blindness in babies.

- Chlamydia often has no symptoms and is treated with antibiotics. It can cause blindness in babies.

- An infected woman can pass a sexually transmitted disease to her baby.

- People who suspect they may have sexually transmitted diseases can be tested at health clinics. They can also receive treatment at health clinics.

Chapter 12 Review

Comprehension: Identifying Facts

On a separate sheet of paper, write the correct word or words from the Word Bank to complete each sentence.

WORD BANK	
abstinence	HIV positive
antibiotics	immune
cancer	Kaposi's sarcoma
chlamydia	sterility
discharge	syphilis
gonorrhea	transfusion
herpes	virus

1) AIDS is a disease of the _____ system.

2) AIDS is caused by a _____ called HIV.

3) A person with AIDS may get a kind of _____ called Kaposi's sarcoma.

4) If a person has AIDS, a blood test will show that the person is _____.

5) _____ is a rare type of cancer that affects the skin or internal organs.

6) A person can receive a _____, in which blood is transferred from one person to another.

7) _____ is the inability to have children.

8) The first symptom of gonorrhea may be a _____.

9) The first symptoms of genital _____ are painful blisters.

10) The disease of _____ can lead to sterility.

11) _____ can cause an eye infection in babies that leads to blindness.

252 Chapter 12 *Preventing AIDS and Sexually Transmitted Diseases*

12) In the second stage of _____, there is a rash, headache, and loss of appetite.

13) Gonorrhea, chlamydia, and syphilis can be treated with _____.

14) Sexually transmitted diseases can be prevented by _____.

Comprehension: Understanding Main Ideas

Write the answers to these questions on a separate sheet of paper. Use complete sentences.

15) List some ways that a person can get AIDS.
16) If someone contracts HIV, how quickly will symptoms of AIDS appear?
17) How are the symptoms of gonorrhea and chlamydia similar?
18) How do health clinics help people who think they may have a sexually transmitted disease?

Critical Thinking: Write Your Opinion

19) How could you show caring and concern for a person who has AIDS?
20) Do you think all pregnant women should be tested for sexually transmitted diseases?

Test Taking Tip If you don't know the answer to a question, put a check beside it and go on. When you have finished the other questions, go back and try the questions with checks again.

Chapter 13

Common Diseases

If you are like most people your age, your health is excellent. And, like most things that are worth doing, keeping your body healthy will take some effort on your part. The decisions you make now about the foods you eat and the habits you form will influence your risk of developing certain diseases.

In this chapter, you will learn about some noncontagious diseases, what causes them, and how they are treated. You will also find out how to reduce your risk of developing some of these diseases.

Goals for Learning

▶ To find out about cardiovascular diseases and what causes them

▶ To identify certain types of cancer and how to prevent them

▶ To explain how asthma affects the respiratory system

▶ To compare and contrast different types of diabetes

▶ To learn about arthritis

▶ To discover what causes epileptic seizures and what to do if someone has a seizure in your presence

Lesson 1

Cardiovascular Disease

Cardiovascular disease
A disease of the heart and blood vessels

Hypertension
High blood pressure

Diseases that affect the heart and blood vessels are called **cardiovascular diseases**. These diseases are common among people in the United States. In fact, they affect one in four people, or more than sixty-eight million Americans. These diseases of the circulatory system are the number one cause of death in the United States. Someone dies from these types of diseases every thirty-two seconds.

What Is High Blood Pressure?

One common cardiovascular disease is high blood pressure, or **hypertension**. Blood presses against the walls of your arteries as your heart pumps blood throughout your body. This force is called blood pressure. Blood pressure usually increases when you exercise or are excited. Blood pressure normally goes down when you sleep or relax. When a person's blood pressure stays high, even at rest, the person has high blood pressure.

Nutrition Tip

Monitor your intake of salt (sodium). You only need 2,400 mg of this mineral each day.

High blood pressure often leads to heart disease because it causes the heart to pump harder than it normally would. High blood pressure also causes the smooth artery walls to become rough. Thus, the heart has to work harder to move the blood through these thickened, irregular channels.

Aerobic exercise and a healthy diet are two ways to reduce the risk of cardiovascular diseases.

Arteriosclerosis
A chronic disease in which the walls of the arteries thicken

Atherosclerosis
A narrowing of the arteries due to a buildup of fat

What Are Some Other Diseases of the Arteries?

At birth, the walls of the arteries are smooth. Over many years, however, fat can collect along these walls. This fat hardens and prevents blood from flowing normally through the arteries. **Arteriosclerosis** is a chronic disease in which the walls of the arteries thicken. The disease also causes the arteries to become rigid. This is why the process that leads to the disease is commonly called "hardening of the arteries."

Arteriosclerosis can cause other diseases. One of these is **atherosclerosis**, which causes large arteries to narrow because of fat buildup along their walls. The buildup narrows the pathway for blood. This, in turn, slows down the flow of blood to the heart. The blood then thickens and could cause blood clots to form. Blood clots block a vessel so much that the amount of oxygen flowing to the heart is greatly decreased. Atherosclerosis can often be prevented by eating a healthy diet low in fats. A blocked artery is shown in Figure 13.1.

Research has shown that many children today have atherosclerosis. Why do you think this is so?

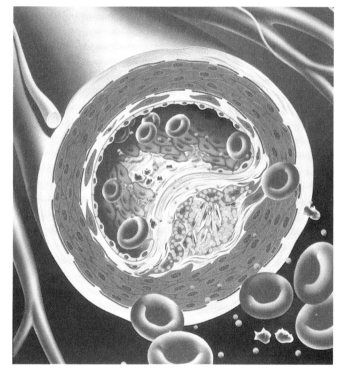

Figure 13.1. This is a cutaway view of an artery that has become blocked with fat.

Heart attack
A condition in which the blood supply to the heart is greatly reduced or stopped

Stroke
A condition in which the blood supply to a person's brain is suddenly blocked

Why do you think heart transplants are rarely used to treat heart attack victims?

What Is a Heart Attack?

Like all the muscles in your body, your heart needs oxygen and other nutrients. A **heart attack** occurs when the supply of blood or other nutrients is greatly reduced or blocked. If a heart attack is mild, only a small amount of tissue is damaged.

People who survive heart attacks often have to take medicine to prevent other attacks. A change in diet, daily exercise, and regular checkups are important to heart attack survivors.

What Is a Stroke?

Another cardiovascular disease is a **stroke**. A stroke occurs when the blood supply to a person's brain is suddenly blocked. Damaged arteries or blood clots can cause strokes. High blood pressure may also cause a stroke.

After a stroke, parts of the body may not be able to function as they once did. Some strokes are mild. In these situations, the victims will notice little change in body functions. Other strokes are severe and may prevent the person from doing certain activities. If the part of the brain that controls speech, for example, is damaged, the person might not be able to speak. Therapy is useful for people who have had strokes.

OCCUPATIONAL THERAPY ASSISTANT

Do you think you would like to help people who have lost some movement of their bodies due to strokes? If so, you might like a career as an occupational therapy assistant. You would work with other medical professionals to help patients perform everyday tasks. Some stroke victims must relearn how to dress themselves, cook, feed themselves, talk, or walk. An occupational therapy assistant also helps patients deal with stress. If the assistant sees that therapy is not helping the person, he or she must suggest ways to change the treatment. There are some one-year training programs for occupational therapy assistants. A two-year degree, or associate degree, however, is usually preferred.

Risk factor
A trait or habit that increases a person's chances of having or getting a disease

What Are the Risk Factors for Cardiovascular Diseases?

A **risk factor** is a trait or a habit that is known to increase a person's chances of having a disease. Some risk factors are inherited and cannot be changed. These factors are called genetic factors. Other risk factors are related to lifestyle choices. These can be changed. The section below lists some genetic and lifestyle risk factors for cardiovascular diseases. Are you or anyone in your family at risk for these diseases?

Some Cardiovascular Risk Factors

- Having family members with cardiovascular diseases increases your chances of having these diseases. Take the time to find out about your family's health history.

- Being male also increases your risk of a heart attack before you reach middle age. After middle age, a woman's risk of heart attacks also increases.

- Aging increases a person's risk of having a heart attack. More that half of all heart attacks occur in people who are 65 or older.

- Race can also increase a person's risk of cardiovascular diseases. African Americans, for example, have a greater rate of hypertension than Caucasians.

Fitness Tip

Do some kind of aerobic exercise every day, especially if cardiovascular disease runs in your family.

LESSON 1 REVIEW Write the answers to these questions on a separate sheet of paper. Use complete sentences.

1) What is hypertension and why can it lead to a heart attack?

2) What is arteriosclerosis?

3) What happens to a person's heart after a mild heart attack?

4) Compare and contrast risk factors that are inherited and those related to lifestyle.

5) How might stress influence risk factors that could lead to cardiovascular diseases?

Lesson 2

Cancer

Benign tumor
A mass of cells that are not harmful

Cancer
A group of diseases marked by the abnormal and harmful growth of cells

Malignant tumor
A mass of cells that are harmful

Metastasize
Spread cancer to distant tissues

Cancer is a group of more than one hundred diseases marked by the abnormal and harmful growth of cells. Cancer is the second leading cause of death in the United States. Cancer can develop in any tissue in the body. Some types of cancer remain in the tissues where they started. Some cancers "travel" in the blood or lymph fluids. They move to other parts of the body that are far from the original site. When cancer spreads to distant tissues, it is said to have **metastasized**.

Some cancers grow and form masses of abnormal cells called tumors. **Benign tumors** are masses of cells that are not harmful. Over 90 percent of all tumors are benign. **Malignant tumors**, on the other hand, are harmful masses that invade normal tissue. They often spread to other organs in the body.

What Are Some Symptoms and Warning Signs of Cancer?

The symptoms of cancer vary according to the type of tissue affected by the disease. A person with lung cancer, for example, might have a cough that doesn't get better with medical treatment. A woman with breast cancer might feel a lump in her breast. Blood in the stool may be a sign of cancer in the intestines or the colon.

A type of cancer might not have a particular symptom. But seven general warning signs could indicate cancer. These are listed in the chart at the left. Look at the letter that begins each phrase. What word do these letters spell?

THE SEVEN WARNING SIGNS OF CANCER

C hange in bowel or bladder habits
A sore that will not heal
U nusual bleeding or discharge
T hickening or lump in the breast or elsewhere
I ndigestion or difficulty in swallowing
O bvious change in a wart or mole
N agging cough or hoarseness

Common Diseases Chapter 13

What Are Some Common Kinds of Cancer?

Chances are you or someone you know either has some type of cancer or will get it. In fact, according to the American Cancer Society, one in three persons has cancer or is at risk of developing the disease. While there are over one hundred different kinds of cancers, some types are common. The chart below shows the cancer incidence by site and sex.

Breast Cancer

One in eight American females will develop breast cancer during her life. Women whose female relatives had or have the disease are at an increased risk of developing breast cancer. Also, women who have a child by age 30 are less likely to develop the disease than childless women. Having children later in life can increase a woman's chances of developing breast cancer.

Prostate Cancer

Prostate cancer is the most common cancer in men. One in eleven American men will get the disease. Most cases of prostate cancer occur in men over the age of 65. In these situations, the disease often develops slowly. Men in their 40s and 50s also get the disease. But it grows more rapidly in men in this age group.

Lung Cancer

The leading cause of death due to cancer in both males and females is lung cancer. What make this statistic so unfortunate is that lung cancer is one of the few cancers that can be prevented. Over 85 percent of people with lung cancer develop the disease because they smoke.

Cancer Incidence by Site and Sex

Men	Women
Prostate Gland 184,500	Breast 178,700
Lung 91,400	Lung 80,100
Colon & Rectum 64,600	Colon & Rectum 67,000
Urinary Bladder 39,500	Uterus 36,100
Non-Hodgkin's Lymphoma 31,100	Ovary 25,400
Skin—Melanoma 24,300	Non-Hodgkin's Lymphoma 24,300
Mouth 20,600	Skin—Melanoma 17,300
Kidney 17,600	Urinary Bladder 14,900
Blood 16,100	Pancreas 14,900
Stomach 14,300	Cervix 13,700
All Sites 627,900	All Sites 600,700

Source: American Cancer Society, Inc., 1998.

Figure 13.2. Different types of cancers are more common in women than in men and vice versa.

Chemotherapy
A cancer treatment that uses drugs to kill cancer cells

Radiation
A type of treatment that uses energy waves to destroy cancer cells

Smoking is a lifestyle choice. Don't give in to peer pressure and start smoking. If you do smoke, even occasionally, stop immediately. If people you know smoke, encourage them to get help to quit.

Skin Cancer

Skin cancer is the most common type of cancer in the United States. About 600,000 new cases of this kind of cancer occur in the United States each year. Most skin cancer is caused by harmful radiation from the sun and the lights used in tanning booths.

How Is Cancer Treated?

There are three basic kinds of treatment for cancer. One common method of cancer treatment is to use drugs that destroy cancer cells. This type of treatment is called **chemotherapy**.

A sunburn you get as a child can cause problems when you become an adult. What does this tell you about the damage that the sun's rays do to skin?

Another form of cancer treatment is **radiation**, which uses energy waves to destroy cancerous tissue. Radiation is often used to treat some forms of thyroid cancer. It is also used to treat small cancerous masses that have not metastasized.

A third treatment for cancer is surgery. Surgery involves removing the cancer cells and any nearby tissue that the cancer may have affected. Surgery is commonly used to treat skin cancers.

CANCER SURVIVAL RATES

In the early 1900s, people with cancer almost always died from the disease or complications related to the disease. Today, however, almost 60 percent of all people under the age of 55 who have cancer will be alive five years after the diagnosis. Although there is no cure for cancer, this statistic is considered a cure. According to the American Cancer Society, more than five million people alive today have had cancer.

Why do you think it is necessary to wear sunscreen even on cloudy days?

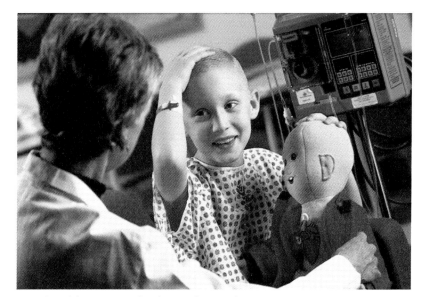

People with cancer who have chemotherapy treatment can lose their hair. This hair loss is often just temporary.

LESSON 2 REVIEW Write the answers to these questions on a separate sheet of paper. Use complete sentences.

1) What is cancer?
2) What is the difference between benign and malignant tumors?
3) What are the seven warning signs of cancer?
4) What causes lung cancer and how can this disease be prevented?
5) Describe the three basic kinds of treatment for cancer.

Suppose a health care plan decides to charge double rates for smokers. Do you think this is fair? Explain your reasons.

Lesson 3

Asthma

Asthma
A disease that affects the lungs, making it difficult to breathe

Asthma is a disease that affects the lungs, making it difficult to breathe. During an asthma attack, a person may cough, wheeze (make a whistle-like sound), and be short of breath. These reactions are due to the swelling and blockage of the airways that carry air from the nose and mouth to the lungs. Asthma attacks can occur as often as once every few hours. Some attacks happen only once every few years.

A person can develop asthma at any age, but the disease commonly affects children. Nearly 12.5 million Americans have asthma. Well over one-third of these people are under the age of 18.

What Triggers Asthma?

Why is exercise a common asthma trigger?

Many things and situations trigger asthma. These triggers are different for different people. They also may affect the same people in different degrees at different times. The box below lists some of the most common asthma triggers.

COMMON ASTHMA TRIGGERS

- Pollen from flowers, trees, and grasses
- Paints and similar chemicals
- Mold spores
- Animal dander
- Household dust
- Cleaning products
- Dirty air filters in heating or cooling vents
- Perfumes, deodorants, cosmetics, and other scented toiletries
- Foods such as eggs, dairy products, fish, and peanuts
- Hard exercise
- Air pollutants
- Cigarette smoke

A person can get asthma at any age, but it often affects children.

How Is Asthma Treated?

Without treatment, asthma can be life threatening. The best way to prevent an attack is to avoid things and situations that trigger asthma. It may be hard to avoid many of the common triggers. But keeping the sleeping area clean and dust-free is an important way to prevent attacks. Putting pillows and mattresses in plastic also helps to prevent severe attacks. Washing bed linens in hot water every week kills dust mites in the linens.

LESSON 3 REVIEW Write the answers to these questions on a separate sheet of paper. Use complete sentences.

1) What is asthma?
2) What happens during an asthma attack?
3) List five common asthma triggers.
4) What is the best way to prevent an asthma attack?
5) What are some things you can do to prevent asthma attacks?

Lesson 4

Diabetes

Type I diabetes
Insulin-dependent diabetes

Diabetes is a group of conditions in which sugar levels are much higher than normal. A person develops diabetes when the pancreas stops producing insulin or doesn't produce enough. Insulin is a hormone that is used by the body to use digested food effectively. Without insulin, the body could not use glucose, a type of sugar, as a source of energy. Glucose would stay in the bloodstream and not be able to enter the cells where it is needed.

About sixteen million people in the United States are thought to have diabetes. Only about half of these people have actually been diagnosed with the condition.

People with diabetes can lead normal lives.

What Is Type I Diabetes?

There are three types of diabetes. **Type I diabetes**, which is also known as insulin-dependent diabetes, is most common in children. The pancreas of a person with this type of diabetes produces little or no insulin. This happens because the body's immune system has destroyed all of the cells that produce this hormone.

One of the most common symptoms of type I diabetes is frequent urination. Other symptoms include increased thirst and hunger. Extreme tiredness and unexplained loss of weight are other symptoms of type I diabetes.

Gestational diabetes
Diabetes that develops during pregnancy

Type II diabetes
Non-insulin-dependent diabetes

How Is Type I Diabetes Monitored?

People with diabetes must always balance their diets, exercise, and insulin to control their blood sugar levels. Blood glucose devices let a person know if his or her blood sugar level is too high or too low. A drop of blood is placed on a special strip. The strip can be read either visually or by a meter. Visual checking involves matching the color of the strip with a standard chart. A person can get a digital reading of his or her blood sugar level with a meter. This method is more accurate than the visual method.

What Is Type II Diabetes?

The second type of diabetes is non-insulin-dependent, or **type II, diabetes**. Between 90 and 95 percent of all cases of the disease are of this type. It is currently not known what causes type II diabetes. But scientists believe that being grossly overweight is the major factor that leads to type II diabetes. Studies show that with this milder form of diabetes, cells "see" the insulin. However, they are unable to "read" its signal to break down the glucose.

When trying to diagnose type II diabetes, a doctor will look at a person's family history to see if diabetes runs in the family. A doctor will also determine if a person's weight might contribute to developing the disease. Thirst and frequent urination may be other symptoms of this type of diabetes. Blood and urine tests are also performed to determine if excess glucose is in these fluids.

What Is Gestational Diabetes?

The third type of diabetes is called **gestational diabetes**. This is because the condition occurs during gestation, or pregnancy. Gestational diabetes often goes away when the pregnancy is finished. The cause of this type of diabetes is related to the nutritional needs of the developing baby. This type of diabetes usually develops about midway through pregnancy. Gestational diabetes is treated by changing the woman's diet. Some women, however, need to use insulin. Women who have had gestational diabetes can develop type II diabetes later in life.

Cataract
A clouding of the lens of the eye

Glaucoma
An eye disease in which pressure damages the main nerve of the eye

What Are Some Health Problems Associated With Diabetes?

Like many diseases, diabetes can cause other health problems. Heart disease is the most common life-threatening disease linked to diabetes. In fact, a person with diabetes has two to four times the risk of developing heart disease. People with diabetes also have a much higher risk of stroke and high blood pressure.

Nerve disease is another health problem common among people with diabetes. About 60 to 70 percent of people with diabetes have mild to severe forms of nerve damage. The damage can cause an impaired feeling in the feet or hands. Severe nerve disease related to diabetes can result in the amputation of the feet or legs.

Diabetes can affect the eyes in many ways. Blurred vision is a common problem among people with diabetes. **Cataracts** and **glaucoma** are eye diseases that occur more often in people with diabetes. Cataracts are a clouding of the lens of the eye. Glaucoma is a disease in which pressure can damage the main nerve of the eye. Blindness can be another complication due to diabetes. Between twelve and twenty-four thousand people with diabetes lose their vision each year in the United States.

LESSON 4 REVIEW Write the answers to these questions on a separate sheet of paper. Use complete sentences.

1) What is diabetes?
2) What causes type I diabetes?
3) What are some symptoms of diabetes?
4) Explain some of the health problems associated with diabetes.
5) What is gestational diabetes? How is it treated?

Lesson 5

Arthritis

Arthritis
A group of diseases marked by swollen and painful joints

Chronic disease
A disease that lasts a long time

Rheumatoid arthritis
A type of arthritis caused by a defect in the immune system

Rheumatoid factor
The antibody associated with rheumatoid arthritis

Arthritis is a group of diseases marked by swollen and painful joints. About forty million Americans have arthritis. Like diabetes and asthma, arthritis is a chronic disease. A **chronic disease** is a disease that lasts for a long time. Most chronic diseases have no cure. With medical care and treatment, however, chronic diseases can be managed.

There are three main types of arthritis. Each has a different cause. One type is thought to be a consequence of aging. Bacteria or viruses cause a second type of arthritis. A third type of arthritis is thought to result from a defect in a person's immune system.

What Is Rheumatoid Arthritis?

Rheumatoid arthritis is a condition caused by a defect in the immune system. The immune system of the affected person reacts to the body's own tissues as if they were foreign. The antibody associated with rheumatoid arthritis is called the **rheumatoid factor**. Once it is activated, inflammation of the connective tissue between the joints occurs. The buildup of scar tissue that replaces the membranes and cartilage in the joints causes pain. The skin, bones, and muscles of the affected joints shrink and wither from disuse and destruction.

What kind of treatment might be used to treat severe cases of arthritis?

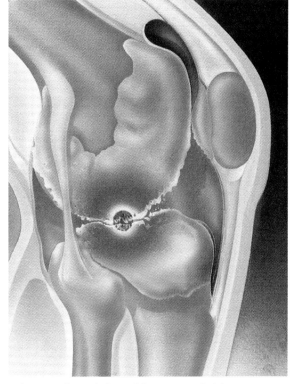

A human knee joint with osteoarthritis.

Osteoarthritis
A type of arthritis that causes a person's joints to get worse with age

Septic arthritis
A swelling of the joints caused by an infection

Rheumatoid arthritis is three times more common in women than in men. The disease progresses gradually in most cases. Pain and stiffness in the joints are followed by swelling and muscle pain. Drugs are often prescribed to relieve pain and control the inflammation. Physical therapy and exercise can also relieve pain and swelling.

What Is Septic Arthritis?

Septic arthritis is a swelling of the joints caused by an infection. Bacteria and viruses can cause joints to swell, feel sore, and fill with pus. Movement can be painful. Treatment of septic arthritis includes resting the affected joints, taking antibacterial and antiviral drugs, and therapy. If treatment is not provided in time, permanent damage to the joints occurs.

What Is Osteoarthritis?

Osteoarthritis is a type of arthritis that causes a person's joints to get worse with age. Both posture and stress on the joints from weight play a major role in the start of the disease. Injury can also lead to this chronic disease.

In the first stages of osteoarthritis, the cartilage in the joints becomes soft. Over time, it is destroyed. Adjacent bones now have no protective covering. This makes the joints stiff, but moving the joints usually relieves the stiffness. Heat and medicines to relieve the inflammation and pain are used to treat this form of arthritis.

Writing About Health

Use what you have learned so far in this chapter to make a list of things to ask your doctor during your next checkup. Make sure you include at least two questions regarding each of these diseases: cardiovascular diseases, cancer, asthma, diabetes, and arthritis.

LESSON 5 REVIEW Write the answers to these questions on a separate sheet of paper. Use complete sentences.

1) What is arthritis?
2) Explain what happens to a person's joints when he or she develops rheumatoid arthritis.
3) What is septic arthritis?
4) What causes osteoarthritis?
5) Name the three types of arthritis and what causes them.

Lesson 6

Epilepsy

Epilepsy
A chronic disease that is caused by disordered brain activity

Grand mal seizure
A seizure that affects a person's motor skills

Seizure
A physical or mental reaction to disordered brain activity

Epilepsy is a chronic disease that is caused by disordered brain activity. Epilepsy affects about 2.5 million people in the United States. The disease occurs more commonly in men than in women. Nearly three-fourths of affected people have their first attack before they are 20 years old.

What Are Epileptic Seizures?

People with epilepsy have seizures. **Seizures** are physical or mental reactions to the brain's disordered activity. During a seizure, a person's mental functions can be disrupted.

A person can become unconscious as a result of a seizure. During some seizures, parts of the person's body can move uncontrollably. Most epileptic seizures last fewer than several minutes. Contrary to popular belief, these seizures don't cause brain damage. Only seizures that last more than half an hour are considered dangerous.

There are different ways that epileptic seizures can be classified. One way is how a person reacts during the seizure. Other ways are where the seizure starts in the brain and the cause of the disordered brain function. There are two main types of epileptic seizures.

What Is a Grand Mal Seizure?

The most common type of epileptic seizure is a **grand mal seizure**. During a typical grand mal, a person suddenly loses consciousness. If the person is standing or sitting, he or she will collapse and fall. The person's muscles will stiffen and relax in a repeating and uncontrollable manner. If the person's rib cage muscles are involved, air is forced from the person's lungs. This causes the person to make grunting sounds.

Sometimes, a person bites his or her tongue during a grand mal seizure because the jaw muscles might stiffen. However, a person cannot swallow his or her tongue during an epileptic seizure.

Petit mal seizure
A seizure that affects a person's mental functions

What Is a Petit Mal Seizure?

A **petit mal seizure** is a less dramatic seizure than a grand mal seizure. Petit mal seizures generally last less than thirty seconds. In most cases, they aren't even noticeable to bystanders. The affected person may simply stare into space for a brief moment. He or she may drop an object that was being held. The person might seem momentarily confused or puzzled. Petit mal seizures originate in a part of the brain that controls thinking. Therefore, thought processes, not motor skills, are disrupted during these types of seizures.

> ### WHAT TO DO IF SOMEONE IS HAVING A GRAND MAL SEIZURE
>
> - Move nearby objects to protect the person from getting hurt by the objects.
> - Loosen the person's clothing.
> - Place a cushion under the person's head.
> - Turn the person's head to the side.
> - *Do not* put any objects into the person's mouth.
> - Call for help if the seizure lasts more than a few minutes.

Children with epilepsy may wear protective equipment to avoid head injuries.

Febrile seizure *A seizure that is common in young children*

What Causes Seizures?

Seizures have many causes. Some well-known causes include tumors, injuries to the head, infections, and diseases that involve blood vessels in the brain. Most seizures, however, have no known cause.

In young children, the most common cause of a seizure is a high fever. Such a seizure is called a **febrile seizure**. These seizures decrease and eventually disappear with age. Febrile seizures are known to run in families. Children who have febrile seizures do not necessarily have epilepsy. A small fraction of these children do develop epilepsy later in life, however.

How Is Epilepsy Diagnosed and Treated?

A diagnosis of epilepsy is often made when a person has had recurrent seizures. A doctor will also examine the person's medical history and conduct complete physical and neurological exams. Laboratory tests on the person's blood and spinal fluid are often used to confirm the disease.

Epilepsy can be successfully treated with medicines. Sometimes, a person will have seizures while taking the medication. Adjusting the amount of medicine usually will bring seizures under control. Once the proper amount of medicine is determined, seizures become less frequent.

Suppose a person is having an epileptic seizure. His legs and arms are twitching. Is he having a petit mal or grand mal seizure? Explain.

LESSON 6 REVIEW Write the answers to these questions on a separate sheet of paper. Use complete sentences.

1) What is epilepsy?
2) What happens during an epileptic seizure?
3) Explain the difference between grand mal and petit mal seizures.
4) What should you do to help a person who is having a seizure?
5) Why is it important to move nearby objects when a person is having a seizure?

An ounce of prevention is worth a pound of cure.
—Proverb

10) Hypertension is another name for high blood _____.

11) When the blood to the brain is cut off, a person may have a _____.

12) Arthritis causes stiff and painful _____.

13) Masses of cells that may be cancerous are _____.

14) A person taking drugs for cancer is getting _____.

Comprehension: Understanding Main Ideas

Write the answers to these questions on a separate sheet of paper. Use complete sentences.

15) How does the body's immune system fight infection?
16) Name five sexually transmitted diseases. Explain which of these diseases are curable and which are not.
17) What are some risk factors for cancer?
18) How is a chronic disease like asthma different from a communicable disease?

Critical Thinking: Write Your Opinion

19) Is it dangerous to be around someone who has AIDS? Explain why or why not.
20) How can someone reduce the risk of having a heart attack?

Unit 5 Review

Comprehension: Identifying Facts

On a separate sheet of paper, write the correct word or words from the Word Bank to complete each sentence.

WORD BANK		
antibodies	infectious	sterility
chemotherapy	inherited	stroke
gonorrhea	joints	tumors
herpes	pathogens	virus
immunity	pressure	

1) Diseases can be either acquired or _____.

2) Germs called _____ cause acquired diseases.

3) The immune system produces _____ to fight infections.

4) A vaccination produces _____ by exposing the body to a dead or weakened virus.

5) Washing hands and cooking food properly can help prevent _____ diseases.

6) AIDS is caused by a _____ called HIV.

7) Some sexually transmitted diseases cause _____, the inability to have children.

8) Painful blisters in the genital area can be a symptom of _____.

9) Antibiotics can treat diseases such as _____ and syphilis.

Unit Summary

- Diseases can be acquired or inherited. Infection, human behaviors, or environmental conditions cause acquired diseases.

- Diseases passed from one person to another are called infectious, contagious, or communicable.

- The germs that cause acquired diseases are pathogens. The body is protected against pathogens through the skin, mucous membranes, stomach acids, coughing, and sneezing.

- Acquired or inherited resistance to disease is called immunity. Vaccinations produce acquired immunity.

- AIDS is a communicable disorder of the immune system caused by the HIV virus.

- AIDS is spread through body fluids during sexual activity, by using infected needles, or from an infected mother to her baby.

- Other sexually transmitted diseases are genital herpes, gonorrhea, chlamydia, and syphilis.

- Cardiovascular diseases such as high blood pressure affect the heart and blood vessels. Heart attacks and strokes are the results of cardiovascular diseases.

- Cancers are diseases marked by the abnormal and harmful growth of cells. Treatments for cancer include chemotherapy, radiation, and surgery.

- Chronic disease affects a person for a long time. Examples of chronic diseases are asthma, diabetes, arthritis, and epilepsy.

- Asthma is a respiratory disease. During an asthma attack, it is difficult to breathe.

- In diabetes, sugar levels are much higher than normal. Some people with diabetes must use insulin because the pancreas doesn't produce enough of this hormone.

- Arthritis results in swollen and painful joints. Rheumatoid arthritis is caused by a defect in the immune system. Septic arthritis is caused by an infection. Osteoarthritis is caused by loss of cartilage.

- Epilepsy is caused by mental reactions to the brain's disordered activity. People with epilepsy have seizures in which they lose control of motor skills.

Deciding for Yourself

Applying Personal Health Habits

You've read a great deal about how diseases are spread. Common infectious diseases may be spread in a number of different ways including:

- through direct physical contact with an infected person
- by droplets coughed into air by an infected person
- by contact with food or water that has been infected by a pathogen
- through the bites of infected animals, such as insects

You can cut down your chances of coming into contact with pathogens or passing them on to others. Of course, it is difficult to stay clear of certain pathogens, but the following steps will help you avoid some contagious diseases.

- Avoid close contact with people who have a cold or the flu.
- Keep your hands clean, and wash them before eating.
- Try not to put your fingers into your eyes or mouth.
- Do not borrow other people's utensils, dishes, toothbrushes, hairbrushes, or makeup.

Questions

1) Why do you think the flu is so easily spread from one person to another?
2) What personal health habits do you do to avoid contagious diseases such as a cold or the flu?
3) Think about how HIV is spread. What can you do to avoid getting HIV?
4) Why do older people who are at nursing homes have a greater risk of getting infectious diseases?

10) _____ is a type of arthritis caused by a defect in the immune system.

11) A disease caused by disordered brain activity is _____.

12) A _____ affects only the body's mental functions.

13) A _____ affects a person's muscles.

Comprehension: Understanding Main Ideas

Write the answers to these questions on a separate sheet of paper. Use complete sentences.

14) How can you prevent cardiovascular diseases?

15) Compare and contrast some common cancers.

16) Distinguish among the three kinds of diabetes.

17) What are the causes of arthritis?

18) Explain the difference between grand mal and petit mal seizures.

Critical Thinking: Write Your Opinion

19) What would you say to encourage someone to stop using tobacco?

20) Are you at risk for developing any of the diseases mentioned in this chapter? What can you do to reduce your risks?

Test Taking Tip

Look for specifics in each question that tell you in what form your answer is to be. For example, some questions ask for a paragraph, and others may require only one sentence.

Chapter 13 Review

Comprehension: Identifying Facts

On a separate sheet of paper, write the correct word or words from the Word Bank to complete each sentence.

WORD BANK		
arthritis	grand mal seizure	stroke
asthma		type I diabetes
atherosclerosis	petit mal seizure	type II diabetes
cancer	rheumatoid arthritis	
diabetes		
epilepsy	seizures	

1) A person with _____ is not dependent on insulin.

2) A group of diseases that cause joints to be stiff and painful is _____.

3) A person with _____ must take insulin.

4) _____ is a disease of the respiratory system.

5) A _____ occurs when blood to the brain is cut off.

6) _____ is a disease in which large arteries narrow because of fat buildup along their walls.

7) Disruptions in the body's physical or mental functions are called _____.

8) _____ is a group of diseases marked by the abnormal and harmful growth of cells.

9) _____ is a group of diseases in which the pancreas produces little or no insulin.

Chapter Summary

- Diseases that affect the heart and blood vessels are cardiovascular diseases.

- High blood pressure, arteriosclerosis, and atherosclerosis are cardiovascular diseases.

- A heart attack occurs when blood to the heart decreases or stops.

- A stroke occurs when the blood supply to the brain is cut off.

- Cancer is a group of about one hundred diseases marked by the abnormal and harmful growth of cells.

- Benign tumors are masses of cells that are not harmful. Malignant tumors are harmful masses of cells that invade normal tissue.

- Cancer is treated with radiation, chemotherapy, or surgery.

- Asthma is a respiratory disease that makes it difficult to breathe.

- Diabetes is a disease in which blood sugar levels are much higher than normal.

- A person with type I diabetes must use insulin. This is because the pancreas produces little or none of this hormone.

- In persons with type II diabetes, body cells "see" the insulin. But they are unable to "read" its signal to break down the glucose.

- Gestational diabetes is a condition that occurs in some pregnant women.

- Rheumatoid arthritis is a condition caused by a defect in the immune system. Septic arthritis is a swelling of the joints caused by an infection. Osteoarthritis causes a person's joints to get worse with age.

- Epilepsy is caused by disordered brain activity.

- During a grand mal seizure, a person loses control of motor skills. During a petit mal seizure, only a person's thought processes are disrupted.

- Seizures can be caused by tumors, head injuries, infections, and diseases of the brain. Most seizures, however, have no known cause.

Unit 6

Injury Prevention and Safety Promotion

Are you prepared for an emergency? Do you know what to do if you or someone in your family has an accident? Knowing how to get help quickly in an emergency is a skill everyone needs. It's even better if you can prevent accidents from happening. But, accidents do happen. Most people have injuries due to accidents at some time.

Many accidents, particularly those in the home, can be prevented by simple safety precautions. In this unit, you will learn ways to help protect yourself and your family from accidental injury or injuries due to violence. You will read about ways to reduce the risk of injury and what you can do if an injury occurs.

▶ Chapter 14 Preventing Injuries
▶ Chapter 15 First Aid for Injuries
▶ Chapter 16 Preventing Violence

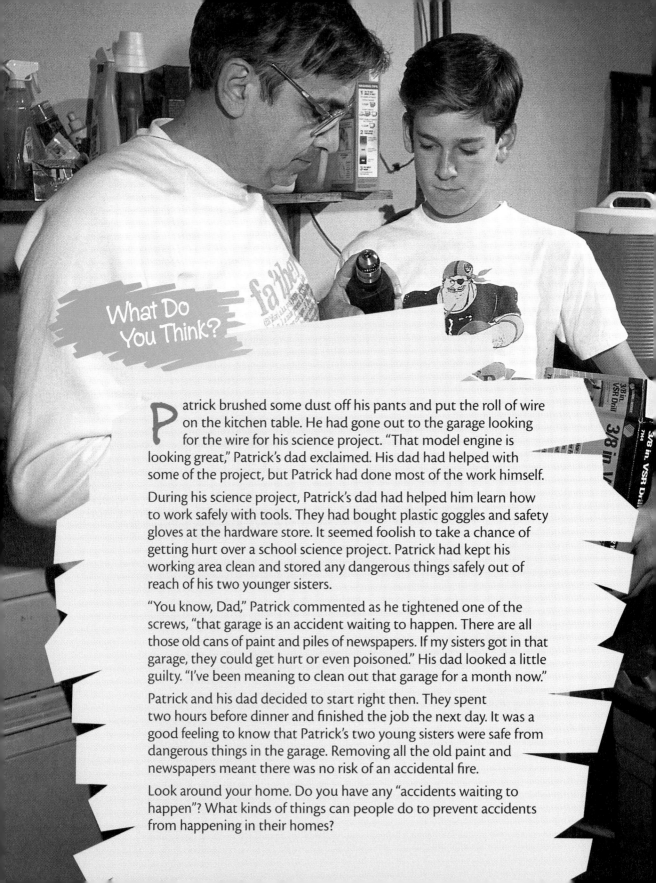

What Do You Think?

Patrick brushed some dust off his pants and put the roll of wire on the kitchen table. He had gone out to the garage looking for the wire for his science project. "That model engine is looking great," Patrick's dad exclaimed. His dad had helped with some of the project, but Patrick had done most of the work himself.

During his science project, Patrick's dad had helped him learn how to work safely with tools. They had bought plastic goggles and safety gloves at the hardware store. It seemed foolish to take a chance of getting hurt over a school science project. Patrick had kept his working area clean and stored any dangerous things safely out of reach of his two younger sisters.

"You know, Dad," Patrick commented as he tightened one of the screws, "that garage is an accident waiting to happen. There are all those old cans of paint and piles of newspapers. If my sisters got in that garage, they could get hurt or even poisoned." His dad looked a little guilty. "I've been meaning to clean out that garage for a month now."

Patrick and his dad decided to start right then. They spent two hours before dinner and finished the job the next day. It was a good feeling to know that Patrick's two young sisters were safe from dangerous things in the garage. Removing all the old paint and newspapers meant there was no risk of an accidental fire.

Look around your home. Do you have any "accidents waiting to happen"? What kinds of things can people do to prevent accidents from happening in their homes?

Chapter 14

Preventing Injuries

Think about your daily life. On a typical day, you probably do many of the same things in pretty much the same order. You wake up. You take a shower, get dressed, and eat breakfast. On weekdays, you go to school, talk with friends, and do homework. Now think about a day that isn't typical. You stay up late to watch a movie and are tired the next day. You miss the bus. You have trouble concentrating. It is on these days that injuries often occur.

In this chapter, you will learn how to avoid injury in certain situations. You will find out what to do to minimize the risks associated with certain lifestyle choices, emergencies, accidents, fire, and some natural disasters.

Goals for Learning

▶ To promote safety by thinking ahead to consider the risks and consequences of certain situations

▶ To learn how to prevent fires and how to react if a fire breaks out

▶ To identify safety rules that pertain to baby-sitting, emergency situations, and the Internet

▶ To identify items that should be in a basic emergency kit

▶ To learn how to prepare for and react during a natural disaster

Lesson 1

Promoting Safety

Are you familiar with the saying "always be prepared"? Simply knowing what to expect or how to react in a situation will prevent many injuries. Most injuries occur because people don't stop to think about the risks they or others might be taking. Too often, people don't think about what could happen in a given situation.

Teenagers are more at risk of injuring themselves than adults are. This is because teens often act suddenly when faced with certain decisions. And, unlike adults, teenagers are less likely to believe that risk is present in certain situations. Think about how your safety might be affected by:

- drugs
- your mood
- being in a moving car
- doing certain sports and activities
- being around firearms

How Can Drugs Affect Safety?

The risk of injury increases when a person uses alcohol or other drugs. Drugs change how a person sees a situation. Drugs can cause the user to take chances that he or she wouldn't normally take. Taking such chances often results in injuries.

Drugs alter a person's reaction time, too. Stimulants, such as caffeine, speed up the function of brain cells. Depressants, such as alcohol and marijuana, decrease the brain's activity. Depressants greatly reduce a person's reaction time and coordination. The number of drug-related injuries in the United States is great. For example, over 50 percent of motor vehicle-related deaths in the United States involve drug-impaired drivers.

What Are Some Motor Vehicle Safety Rules?

In 1998 alone, nearly 6,000 teenagers died from injuries suffered in motor vehicle crashes. In the same year, another 350,000 teenagers were injured in such collisions. As a passenger, you can reduce the chances of a collision by following these rules.

- Keep the noise levels down so that the driver can concentrate.
- Refuse to get into a car with a driver who has been drinking alcohol or taking other kinds of drugs.
- Ride only with drivers who obey all traffic laws.

If a crash does occur, injuries and deaths can be greatly reduced if you obey the rules below.

- Wear a seatbelt and remind others to wear theirs *at all times*.
- Keep your hands, head, legs, and feet inside the car while it is moving.
- Ride in the front seat of a car with air bags only if you meet the weight and height limit.

Writing About Health

Although statistics show that seat belts save lives, many people refuse to wear them. Write a persuasive paragraph to convince people to wear seat belts.

What are some other ways to prevent car accidents?

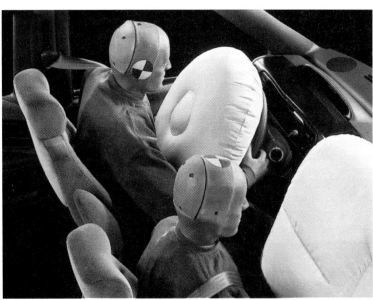

Air bags at work in an automobile.

Preventing Injuries Chapter 14

How Can You Prevent Certain Sports-Related Injuries?

As you probably know, sports and other forms of physical activity are good ways to stay healthy and fit. These activities also help you to relax and to relieve stress.

There are many ways to prevent sports-related injuries. Some of these are listed below.

What kinds of safety equipment should be worn to play soccer?

- Make sure you have the strength and endurance for a sport or an activity.
- Condition and warm up before participating in a game or an activity. Cool down after the game or activity.
- Wear proper safety equipment including a face mask, mouth guard, shin guards, goggles, and an appropriate helmet.
- Make sure your shoes and clothing fit properly. Check laces, buckles, and straps often for signs of wear.

How might you protect yourself from serious injury if you fall while skating?

- Play or carry out the activity at a well-maintained facility or in an area appropriate to the sport or activity.
- Always drink water before, during, and after the game or activity. This prevents heat exhaustion and dehydration.

SAFER CARS

The first automobiles had no doors, roofs, or windshields. They obviously weren't as safe as modern cars. Of course, they weren't as fast as today's vehicles, either. Traffic at that time doesn't compare to that on today's roads. Over the years, the number of cars has greatly increased. So have driving speed and concerns about safety. Seat belts, antilock brakes, safety glass, front-wheel drive, brake and signal lights, and shock-absorbing bumpers are just a few of the safety features of most cars on the road today. In the early 1990s, air bags became standard equipment in new cars. Air bags, when properly used, have reduced injuries from front-end collisions and have also reduced the number of traffic fatalities.

INDUSTRIAL SAFETY SPECIALIST

A workplace should be safe for the people who work there. An industrial safety specialist is a person who works to make sure companies identify and correct unsafe working conditions. The duties and responsibilities of these specialists are to prevent work-related injuries and deaths and to protect the health of all workers. Industrial safety specialists must be concerned with waste disposal, environmental safety, pollution, and health hazards. Most industrial safety specialists have a college degree. Continuing education is also necessary to keep up with changes in laws that regulate workplace safety.

Firearm
A handgun or rifle

What Can You Do to Be Safe Around Firearms?

Firearms, or handguns and rifles, are in about half of all homes in the United States. Most firearm injuries and deaths take place inside a house or an apartment or close to the buildings. The majority of firearm accidents among young people occur because children are curious about firearms. Another reason people are injured is that they don't realize the power of a handgun or rifle.

To reduce the risk of a firearm accident, handguns and rifles must be stored in locked containers or cabinets. The ammunition for the firearms must also be in a locked container that is separate from the firearms.

LESSON 1 REVIEW Write the answers to these questions on a separate sheet of paper. Use complete sentences.

1) Why are teenagers more at risk of injuring themselves than adults are?

2) What are two ways in which drugs affect safety?

3) Why is it important to drink water before, during, and after a game or an activity?

4) What are some rules to follow to reduce the chances of injury from a car crash?

5) How can you reduce the chances of injury from firearms?

Lesson 2

Reducing Risks of Fire

Hazard
A danger

House fires are among the leading causes of serious injuries and deaths in the United States. This is sad because over 90 percent of these fires are preventable. Most house fires are due to carelessness. The number of house fires could be greatly reduced. People only need to take care when they cook, smoke, or burn wood in their fireplaces. Checking for and fixing faulty wires and electrical cords can also help to reduce the risk of a house fire. Another way to prevent fires is to keep matches, cigarette lighters, and fireworks out of children's reach.

How Can You Reduce the Risks of House Fires?

Take an active role in reducing the risk of fire in your house or apartment. With a responsible adult, look for possible fire dangers, or **hazards**. Use the list below as a guide. Can you think of any other possible hazards?

- Look for damaged wires and electrical cords.
- Don't overload electrical outlets or extension cords.
- Make sure all appliances are in good working order.
- Store flammable liquids and household cleaners away from all heat sources.
- Store gasoline only in an approved container.
- Make sure that matches and lighters are well out of reach of children.
- Place a fire screen in front of a fireplace when a fire is burning.
- Test smoke detectors twice a year to make sure they are working properly. Change the batteries two times a year.
- Turn pot and pan handles toward the center of the stove when cooking. Take care not to get cooking oil and grease too hot.

Safety Tip

If a door is closed and you suspect a fire, feel the door and doorknob with the back of your hand to see if they are warm. A warm door or doorknob could signal fire on the other side. If the door or knob is warm, choose an alternate escape route.

Why should there be at least two neighbors that you can go to during a fire?

Smoke detectors should be installed in hallways, bedrooms, the basement, and the kitchen.

What Should You Do If a Fire Breaks Out?

Even if you have fireproofed your house or apartment, fires can happen. It is important to know that a fire will usually burn for two to four minutes before a smoke detector goes off. The size of a fire increases eight times every minute. That means you only have about a minute to get out of a burning structure. Therefore, it is important to get out quickly and then call 911. There are a few situations in which you might try to extinguish, or put out, the fire. First, if the fire is small and hasn't yet spread, you could try to put it out. Second, if a family member has his or her back to an escape route, an attempt to put the fire out could be made. Third, a family member could put out the fire if he or she uses a fire extinguisher correctly.

Sometimes you know that the fire can't be extinguished. Then it is important that everyone leave the house or apartment building immediately. Make sure no one stops to take any belongings. Be sure that everyone follows one of the predetermined escape routes. Go directly to the meeting place. Check in with an adult in the group. Go to a neighbor who is not at risk from the fire and call the fire department or 911.

Action for Health

PLAN FIRE ESCAPE ROUTES

Do you and other members of your family know how to get out of the house or apartment in case of a fire? Use these guidelines to make a plan of escape from *every* room in your home.

- Draw a floor plan of your house or apartment. Show at least two escape routes from each room, such as a door and a window.
- Practice crawling along the floor toward the escape exits.
- Make sure everyone knows where the outside emergency meeting place is.
- Decide who will help people in the family with special needs.
- Post the floor plan where everyone can review it once a month. Quiz family members to make sure everyone knows the escape routes and the location of the emergency meeting place.
- Beneath the floor plan, write the names and addresses of two neighbors to whom you can go in the event of any emergency.

Why should you crawl along the floor to escape a fire?

LESSON 2 REVIEW Write the answers to these questions on a separate sheet of paper. Use complete sentences.

1) Why are most house fires preventable?
2) Explain the importance of having a fire extinguisher and knowing how to use it.
3) What kinds of things should be included in a fire drill plan?
4) What should you do if you are in a building where a fire has started?
5) Why should you *never* reenter a burning building?

Lesson 3

Safety for Teens

Internet
The worldwide computer network that provides information to users

In many situations, teens are more at risk of injuring themselves or causing others to become injured. This is because of their inexperience with the situations. You can learn ways to reduce and prevent injuries while baby-sitting. You can find out what to do in an emergency situation and how to protect yourself when you are walking alone or are at home by yourself. There are also safety rules to follow when using the **Internet**. The network of computers that provides information is the Internet.

SAFETY FOR BABY-SITTERS

Baby-sitting or watching young children carries many responsibilities for safety. Here are some guidelines that can help avoid risks when you are watching young children.

- Know the number where parents can be reached and know when they will return.
- Have emergency and neighbors' phone numbers.
- Learn family rules for playing inside.
- Learn how to raise or lower the sides of a crib.
- Keep full attention on the children.
- Keep children away from appliances, matches, cleansers, soap, medicine, and bodies of water.

This chart lists some of the rules to follow when baby-sitting.

What is the most important rule to follow when baby-sitting?

What Are Some Rules to Follow When Baby-sitting?

Baby-sitting can be fun and a good way to earn spending money. But because children are unaware of the consequences of many of their actions, watching them demands your full attention.

Choking, poisoning, and drowning are the three leading causes of death for children under age five. With this in mind, always keep the children you are watching away from small objects. Items like marbles, coins, and hard candy could cause choking. Cut their food into small bite-sized pieces. Don't let children run around while they are eating.

As a responsible baby-sitter, you must keep children away from bathtubs, toilets, buckets, and swimming pools. Children can drown in less than an inch of water. Supervise bath time for young children. Never leave them unattended for even a moment.

Life-threatening emergency
Any situation in which a person might die if medical treatment isn't provided immediately

Who Should You Call in an Emergency?

In this chapter, you've learned about different kinds of injuries and how you might prevent them. In spite of precautions, however, injuries still occur. Thus, you should be aware of what to do when a person is injured.

Some injuries, as you might know from personal experience, are not serious. Minor burns, some insect and animal bites, and sprains are injuries that might require medical attention. But usually they are not **life-threatening emergencies**. A life-threatening emergency is a situation in which a person could die if treatment isn't immediately available. Shock, choking, stroke, heart attacks, poisoning, and severe bleeding are life-threatening emergencies.

Suppose you and a friend are riding your bikes. Your friend falls and scrapes her leg badly. Who should you call? Parents or other responsible adults, such as neighbors and relatives who live close by, can be contacted for help.

Now, suppose a passing car hits your friend. She is unable to move. In this situation, you or another person would call 911, a central emergency center. When you dial this number, people are immediately available to provide help. When calling 911, stay calm and talk slowly. Provide clear information about the situation. Answer all of the questions you are asked. Stay on the line until you are told to hang up. And remember, call 911 only if a real emergency exists.

A CALL FOR HELP

The **911** telephone number is for emergencies. The system is available in about half of the United States. Where the system isn't available, calls can be made to an emergency operator, the police, or the fire department. The caller can help the 911 operator by providing information calmly, answering all questions, and following directions until instructed to hang up.

How Can You Protect Yourself From Being Hurt by Others?

If you are like many teens today, you might be alone at home before or after school. You might have to walk or bike to and from school alone. If you are in these or similar situations, follow these rules to reduce your chances of being hurt by others.

- Find classmates to accompany you if you walk or ride your bike to and from school. If this isn't possible, walk or ride your bike only on well-traveled streets. Avoid shortcuts, small side streets, alleys, and wooded areas.
- Do not accept rides from strangers, even if they seem nice and appear to want to help you.
- If you are being followed by a stranger, quickly go to a crowded street, a public building, or some other active area.
- When you get home, call a parent or other adult immediately to say that you have arrived safely.
- When you are home alone, make sure you don't tell unknown phone callers that you are alone. Decide with the adults in your family what you should say if someone you don't know calls. Practice the script and keep a copy of it near the phone.
- Keep the door locked and do not open it. If someone comes to the door, speak through the closed, locked door—even if you know the person.
- Make sure a list of emergency numbers is by each phone in your house or apartment.

How Can You Protect Yourself on the Internet?

You have probably discovered that the Internet is a great place to "hang out." You can do research on the Internet to help you complete school projects. You can use this powerful tool to visit sites to find books by a certain author.

E-mail
Messages sent and received over the Internet

You can view online catalogs of your favorite clothes. You can listen to music and find tips and codes for video games. You can send electronic messages called **e-mail** to friends and family. You can even chat with people next door, across town, or in another country.

While the Internet is a useful tool, you must realize that some people can harm you. Follow these rules to make using the Internet a safe and enjoyable activity.

- Avoid any sites that contain material or activities that make you uncomfortable. If you accidentally go to such a site, leave it immediately by clicking on your Home icon.

- Never give out any information about yourself, your family, or where you live to people or sites without first talking to a parent or other responsible adult.

- Don't go into chat rooms unless you have permission from a parent or guardian.

How is the Internet like a big city?

- If you post something at a site, never include your name, home address, or phone number.

- Never respond to e-mail from people you don't know.

- Do not agree to meet face-to-face with someone you meet online unless you have permission from a parent or guardian.

LESSON 3 REVIEW Write the answers to these questions on a separate sheet of paper. Use complete sentences.

1) What are some important things to remember when you are baby-sitting?

2) What are life-threatening emergencies?

3) When calling 911, what should you do?

4) If you think a stranger is following you, what should you do?

5) What are some rules to follow when using the Internet?

Lesson 4

Emergency Equipment

Emergency kit
A collection of items that are useful in almost any kind of emergency

Suppose you are baby-sitting and a bee stings one of the children. Do you know where to find the things you need to care for the bite? Suppose you are at home alone in the evening and the power goes out. Do you know where the flashlight is? You need to know about emergency equipment that should always be on hand in any house or apartment.

What Kinds of Things Might You Need in an Emergency?

An **emergency kit** is a collection of items that can be used in almost any kind of emergency. Read about these different items.

Radio

A battery-powered radio is an important item in an emergency kit. It could be your only source of information when the power in your area is out. Make sure your family's emergency kit contains a portable radio and the batteries to run it. In a citywide or statewide emergency, certain radio stations report and update people on the situation. They tell you how you and your family should respond.

Sources of Light

During some emergencies, people can be without electricity for hours, days, or even weeks. At least two flashlights should be available for power outages. Packs of fresh batteries should be kept near the flashlights.

Candles are also good sources of light in an emergency. Make sure the candles are in sturdy candleholders so that they are not likely to tip over. Keep matches to light the candles in a heavy plastic bag near the candles. Never light a match, however, if you smell gas or suspect a gas leak.

Preventing Injuries Chapter 14

These are some of the emergency supplies that you should always have on hand.

First Aid Kit

A good first aid kit contains bandages, ice packs, scissors, tweezers, gauze, tape, and medicines. These items can be used to treat simple injuries. They can also be used to treat serious injuries until professional help arrives. Check your first aid kit after each emergency. Replace items that were used or those that have expired.

LESSON 4 REVIEW Write the answers to these questions on a separate sheet of paper. Use complete sentences.

1) What is an emergency kit?
2) Why is a battery-powered radio an important item in an emergency kit?
3) Why shouldn't you light a match if you smell gas?
4) Name some items a first aid kit should contain.
5) Why is it important to have both candles and flashlights in an emergency kit?

What are some items specific to your family that you would include in an emergency kit?

Lesson 5
Safety During Natural Disasters

Natural disaster
A destructive event that happens because of natural causes

Earthquake
A shaking of the rocks that make up the earth's crust

Natural disasters are destructive events that happen because of natural causes. Some natural disasters, such as earthquakes and hurricanes, occur only in certain regions of the United States. Other disasters, such as floods and thunderstorms, can happen in all parts of the country. Do you know what to do to prepare for a natural disaster? What should you do during and after the disaster?

How Should You Prepare for an Earthquake?

An **earthquake** is a shaking of the rocks that make up the earth's crust, or outer layer. Earthquakes are caused when blocks of the earth's crust and middle layer shift. Most earthquakes in the United States occur along the west coast. However, the most destructive earthquake in the history of the United States took place in Missouri. Therefore, everyone should know how to prepare for an earthquake as well as how to react during this natural disaster.

If you live in a place that has earthquakes often, securely bolt heavy objects, like bookshelves, to a solid wall. Also make sure that heavy items are stored on shelves close to the floor. Don't hang heavy picture frames or mirrors above beds.

Locate the shutoff valves of the gas line and water main. Also know how to turn off the power in your house or apartment. Have a responsible adult show you how to turn off these valves.

Safety officials often stage natural disasters so that they can learn how to react to these events.

Preventing Injuries Chapter 14

Earthquakes can cause buildings and highways to topple and crumble. Water lines and gas lines are also damaged during earthquakes. Broken gas lines are the most common cause of fires related to earthquakes.

What Should You Do During and After an Earthquake?

During an earthquake, try to stay calm. If you are inside, get under a sturdy table or desk. Crouch down and cover your head and face with your arms. If the table or desk starts to move, hold onto it and move with it. Stay where you are until the shaking stops completely. Stay away from windows and glass doors. Don't try to run outside or use an elevator, an escalator, or the stairs.

If you are outdoors, stay in an open area. Avoid buildings, power lines, trees, or other objects that could topple during the earthquake. When it is safe to move about, make sure that you stay away from downed power lines. They could kill you if you touch them.

Hurricane
A tropical storm that forms over the ocean

Hurricane warning
A situation in which a hurricane has reached land

Hurricane watch
A warning to prepare for a hurricane

Thunderstorm
A severe weather condition that produces thunder, lightning, and rain

If you are with a group of people outdoors during a thunderstorm, is it better to spread out or stay together? Explain your answer.

Once the shaking stops, clean up any spills and broken glass. Have an adult check for gas and water leaks and for broken electrical wires. If damage is found, remind the adult to turn off the water, gas, or electricity at the source. Check the inside and outside of your house or apartment for cracks and other damage to walls.

How Can You Stay Safe During a Thunderstorm?

A **thunderstorm** can form anywhere in the United States. It is a storm that produces thunder, lightning, and large amounts of rain. While thunder can be scary at times, it cannot harm you. Lightning, however, is dangerous. A single lightning bolt can discharge millions of volts of electricity. When indoors during a thunderstorm, stay away from open doors and windows. Don't use electrical appliances. Stay away from sinks and tubs that contain water. Do not use the phone except in an emergency.

If you are outdoors, avoid tall trees, hilltops, and isolated metal objects, such as fences and water towers. Crouch down and bend over to make yourself as small as possible. Do *not* lie flat on the ground and do *not* lie in a low spot where water can collect.

What Dangers Are Associated With Hurricanes?

Hurricanes are tropical storms that form over the ocean. The winds generated by these storms blow at speeds of at least 75 miles per hour around a storm center called an eye. Hurricanes generally affect the eastern United States and states that border the Gulf of Mexico.

Hurricanes are storms that are easily tracked. If a hurricane is expected to reach land within a few days, a **hurricane watch** is issued. During a watch, people in the affected area should prepare for the heavy winds and rains produced by the storm. When a **hurricane warning** is issued, people should evacuate, or leave, the area. Before leaving, the gas and water mains should be closed. Electricity should also be turned off. Only necessities should be taken to the evacuation points. People should stay at the emergency shelters until officials tell them to leave.

Flood
A condition in which a body of water overflows and covers land that is not usually under water

Flood warning
A situation in which flooding has occurred

Flood watch
A situation in which flooding is possible

Tornado
A whirling, funnel-shaped storm that forms over land

Tornado warning
An alert issued when a tornado has been spotted in an area

Tornado watch
A situation in which a tornado may develop

Safety Tip

Find out what kinds of natural disasters are likely to affect your area. With an adult, make plans that you can follow during each type of disaster.

How Can You Be Safe During a Tornado?

A **tornado** is a whirling, funnel-shaped storm that forms and moves over land. When a **tornado watch** is issued, there is a possibility that a tornado will develop. A **tornado warning** is issued when a tornado is spotted in an area.

When a tornado warning is issued in your area, follow these rules. If indoors, go to the basement and stay there. If the building you are in doesn't have a basement, go to the lowest floor. If you are outdoors during a tornado warning, lie flat in a low spot on the ground. If you are in a car, get out and go into a building if possible. If you are in the storm's path, move away from the funnel cloud at a right angle to its path.

What Should You Do If Floods Affect Your Area?

A **flood** is a condition in which a body of water overflows and covers land that is not usually under water. If the possibility of flooding exists, a **flood watch** is issued. A **flood warning** is a situation in which you should head for higher ground because flooding has occurred.

During a flood, don't try to cross a stream in which water is at or above knee level. If you are a passenger in a car, make sure the driver does not drive through an underpass that is flooded. If the car stalls, make sure all the passengers get out and go to a safe place.

LESSON 5 REVIEW Write the answers to these questions on a separate sheet of paper. Use complete sentences.

1) If you are outdoors during an earthquake, what should you do?
2) If you are indoors during a thunderstorm, what should you do?
3) What should you do if a hurricane warning is issued?
4) If you are outdoors during a flood, what should you do?
5) Suppose that during a severe thunderstorm, you are outside surrounded by tall trees. How would you protect yourself?

Chapter Summary

- Teens are more at risk of injuries than adults. This is because teens often act suddenly when faced with certain situations.

- Drugs increase a person's risk of injury because drugs change one's perception and alter reaction time.

- Keeping noise levels down so that the driver can concentrate can reduce car collisions.

- Warming up and cooling down before and after a game or an activity reduces your chances of injury.

- To reduce firearm injuries, always treat a handgun or rifle as if it is loaded.

- Nearly all house fires are preventable. Checking the house for fire hazards reduces the number of fires. Knowing what to do if a fire breaks out reduces your risk of injury.

- A responsible baby-sitter keeps full attention on the children. A responsible sitter also knows what to do and who to contact in an emergency.

- When calling 911, stay calm, talk slowly to the operator, and don't hang up until you are told to do so.

- Use rules to make using the Internet a safe and enjoyable activity.

- An emergency kit should include a battery-operated radio, flashlights, batteries, candles, matches, and a first aid kit.

- Natural disasters can be caused by earthquakes, tornadoes, hurricanes, floods, and thunderstorms.

Chapter 14 Review

Comprehension: Identifying Facts

On a separate sheet of paper, write the correct word or words from the Word Bank to complete each sentence.

WORD BANK		
earthquake	flood warning	thunderstorm
e-mail	hurricanes	tornado
emergency kit	hurricane watch	tornado watch
firearms	Internet	
flood	life-threatening emergency	

1) _____ are storms that form over oceans.

2) You can use a computer to find information on the _____.

3) The shaking of the rocks that make up the earth's crust is an _____.

4) _____ is an electronic message sent over the Internet.

5) A severe storm that produces thunder, lightning, and lots of rain is a _____.

6) A _____ is issued when rivers overflow their banks and cover surrounding land.

7) Handguns and rifles are _____.

8) A _____ occurs when land that is normally dry is covered with water from a river or stream.

9) A _____ is issued when a tropical storm may hit nearby land.

10) A person having a heart attack is an example of a _____.

11) Flashlights, a radio, and a first-aid kit are important items that should be in an _____.

12) When weather conditions are such that a funnel-shaped storm might form, a _____ is issued.

13) A _____ is a funnel-shaped cloud that forms on land and can cause much destruction as it moves in a narrow path.

Comprehension: Understanding Main Ideas

Write the answers to these questions on a separate sheet of paper. Use complete sentences.

14) Why do most injuries occur?

15) What should be included in a fire escape plan?

16) How can you protect yourself from danger while walking home alone?

17) Name at least three things specific to your family that you would include in an emergency kit.

Critical Thinking: Write Your Opinion

18) Why do you think everyone should be prepared for each kind of natural disaster?

19) How do you protect yourself against harm when using the Internet?

20) What would you do if a friend drank some alcohol and insisted on driving?

Test Taking Tip — Avoid waiting until the night before a test to study. Plan your study time so that you can get a good night's sleep the night before a test.

Chapter 15

First Aid for Injuries

Have you ever accidentally cut yourself? Did you ever take a bad fall off a bike or see a friend sprain an ankle? Each of these situations is an emergency—a sudden need for quick action. An emergency can happen at any time. Knowing how to handle it can keep a bad situation from getting worse. It might mean the difference between life and death.

In this chapter, you will learn how to provide basic first aid in emergencies. You will learn the signs of choking, respiratory failure, cardiovascular failure, severe bleeding, and shock. You will learn how to provide first aid for these and other life-threatening emergencies.

Goals for Learning

▶ To list some of the guidelines for basic first aid
▶ To describe basic first aid techniques for common injuries
▶ To demonstrate the Heimlich maneuver and describe the universal sign for choking
▶ To explain the steps of rescue breathing and CPR

Lesson 1

What to Do First

First aid is the immediate emergency care given to a sick or an injured person before professional medical help arrives. The information in this chapter is only a first step toward first aid training. Classes offered by schools, fire departments, and park districts help you develop your first aid skills.

What Are the Basic Guidelines for First Aid?

The most important thing to do for a victim in an emergency is to remain calm. This helps you think clearly and provide the best care possible. Then follow these guidelines:

1. Look around the immediate area.
   Are you or the victim in any danger of fire, explosion, or drowning? If so, move the victim and yourself out of harm's way. If there is no danger, do not move the victim.

2. Find out if the victim is conscious by tapping the shoulder or loudly asking if he or she is all right. If the victim is not conscious, stay there and ask someone to call for help. If you are alone, quickly call 911 or 0 for the operator. This will put you in touch with the **Emergency Medical Service (EMS)** for your area. EMS personnel will ask you for your name, location, and details about the emergency.

3. Quickly return to the victim and check for breathing and a pulse. If the person is conscious, ask permission to provide care. Check for any injuries. See if the person has emergency medical identification, such as a tag, bracelet, or card. Cover the victim with a blanket or coat to prevent shock. Apply direct pressure to bleeding areas. Continue to provide care until help arrives.

> **Emergency Medical Service (EMS)**
> *An intercommunity emergency system that sends out fire, police, and ambulances by dialing 911 or 0 for operator*
>
> **First aid**
> *The immediate care given to a sick or an injured person before professional help arrives*

Safety Tip
Keep a list of emergency numbers near your phone.

Emergency Medical Service staff provide first aid for injured people before moving them to a medical facility.

First Aid for Injuries Chapter 15 **305**

technology

911 SOFTWARE

Every day thousands of emergencies arise, and people dial 911 for fire, police, or EMS help. This life-saving emergency software has caller identification and a dispatch system that sends the appropriate emergency help quickly. The dispatcher's screen shows the caller's address and phone number before the call is answered. Computerized maps connect to the database to provide the quickest route to the caller. Deaf and hard-of-hearing people can call 911 by using a TDD (telecommunications device for the deaf). All calls are recorded.

Good Samaritan Laws
The laws that protect people who assist victims in an emergency

Infectious
Contagious

Universal Precautions
The methods of self-protection that prevent contact with blood or other body fluids

What Are Good Samaritan Laws?

Many states have **Good Samaritan Laws**. These laws protect people who provide first aid in an emergency. These people must use common sense and the skills for which they are trained. The American Red Cross and the American Heart Association provide classes in basic first aid. Rescuers are not expected to risk their own lives in providing emergency care.

What Are Universal Precautions?

A person who provides first aid should use **Universal Precautions**. These precautions protect rescuers or victims from **infectious**, or contagious, disease by preventing contact with blood or other body fluids. Universal Precautions include wearing latex gloves and masks whenever possible. Another Universal Precaution is disposing of materials like bandages that have come in contact with body fluids.

LESSON 1 REVIEW Write the answers to these questions on a separate sheet of paper. Use complete sentences.

How would you help a friend who broke an ankle while playing soccer?

1) What is first aid?
2) Why is it important to remain calm in an emergency?
3) What are Good Samaritan Laws?
4) Provide three examples of Universal Precautions.
5) Suppose you see an older adult lying on the sidewalk. He had been shoveling snow. What would you do?

Lesson 2

Caring for Common Injuries

Elevate
Raise

Fracture
A cracked or broken bone

Splint
A rigid object that keeps a broken limb in place

Sprain
The sudden tearing or stretching of tendons or ligaments

Most situations that require first aid are slight to average emergencies. You have probably experienced some of the injuries described in this lesson. Were they treated correctly? Proper immediate care can prevent an injury from leading to other problems.

What Is First Aid for Broken Bones and Sprains?

If the injured part of the body is swollen, crooked, or bruised, the person may have a **fracture**, or broken bone. He or she may be in severe pain or be unable to move that body part. Call EMS immediately. Keep the body part in the same position in which you found it and apply a **splint**. A splint is a rigid object that keeps a broken limb in place. It can be made of wooden sticks, rulers, pencils, or rolled up magazines. You should tie the splint above and below the fracture to keep that part of the body from moving.

A **sprain** is the sudden tearing or stretching of tendons or ligaments connecting joints. Sprains usually occur in ankles, wrists, or knees. Sprains often occur while playing sports but can also happen while walking. If you suspect a sprain or fracture, remember to ICE (immobilize, cold, elevate) it. First, immobilize the limb. Next, apply cold compresses to reduce swelling. Then, **elevate** the limb to reduce further swelling.

What would you do if you are walking with a friend and your friend sprains an ankle?

For first- and second-degree burns, run cold water to cool the burn.

First Aid for Injuries Chapter 15

Healthy Subjects: Social Studies

THE AMERICAN RED CROSS

The Red Cross began in Switzerland in 1863. A nurse named Clara Barton helped start the American Red Cross in the United States in 1881. The Red Cross helps people during wars, national disasters, or other serious needs. The Red Cross relies on volunteers to help with direct care, fund-raising, and disaster relief. They donate blood used by hospitals during national emergencies. Volunteers make the Red Cross work.

What Other Problems Require First Aid?

Burns

There are three types, or degrees, of burns. A first-degree burn damages only the outer layer of skin. Most sunburns are first-degree burns. A second-degree burn affects the outer and underneath layers of skin. The area may swell and blister. A third-degree burn extends through all layers of skin to the tissues underneath.

For first- and second-degree burns, stop the burn by removing the source of heat. Then cool the burn by applying large amounts of cool water. Finally, cover the area loosely with clean bandages to prevent infection. Third-degree burns require immediate professional attention. Call EMS, stay with the victim, and watch closely for signs of shock. You will learn about shock later in this chapter.

Objects in the Eye

If there is an eye injury where you suspect an object is still in the eye, do not rub the eye. Flush it with lukewarm water starting from the corner near the nose. This may remove the object. If not, seek medical help.

Writing About Health

Think about outdoor activities you enjoy throughout the year. Write about these activities. Tell the ways you can protect yourself from injuries related to temperature.

Heat exhaustion
A condition resulting from physical exertion in very hot temperatures

Heatstroke
A condition resulting from being in high heat too long

Hypothermia
A serious loss of body heat resulting from being exposed to severe cold temperatures

Nosebleeds

You or someone you know has probably had a nosebleed. For a minor nosebleed, sit down and lean forward. Hold the nostril with direct pressure for about ten minutes while breathing through your mouth. You can also apply a cold cloth across the forehead or bridge of the nose. If the bleeding doesn't stop, seek medical help.

Exposure to Heat and Cold

Heat exhaustion occurs from too much activity or physical exercise in very hot temperatures. Signs of heat exhaustion include weakness, heavy sweating, muscle cramping, headaches, and dizziness. To give first aid for heat exhaustion, move the person out of the heat. Loosen the clothing, apply cool cloths to the forehead and neck, and offer water to sip.

Heatstroke may result when a person stays in the heat too long. The main sign of heatstroke is a lack of sweating. Other signs include red skin, vomiting, rapid pulse, confusion, very high body temperature, and sudden unconsciousness. Call EMS or 911 immediately and treat the same as for heat exhaustion.

The opposite of heatstroke is **hypothermia**, or being cold too long. This can occur if someone is not dressed properly in cold weather or if clothing becomes wet. Signs of hypothermia include shivering, slurred speech, and below-normal body temperature. A person can die from hypothermia. If the person isn't breathing, perform rescue breathing. Have someone call for help immediately.

Action for Health

FIRST AID TRAINING

One of the best ways you can prepare for emergencies is to take a class in first aid training. Instructors will teach you the skills you need to give effective first aid. Through these classes, you will practice different first aid treatments on other members of the class and on life-size dolls. Check with your fire department, hospital, community center, park district, Red Cross, or American Heart Association for information about first aid classes.

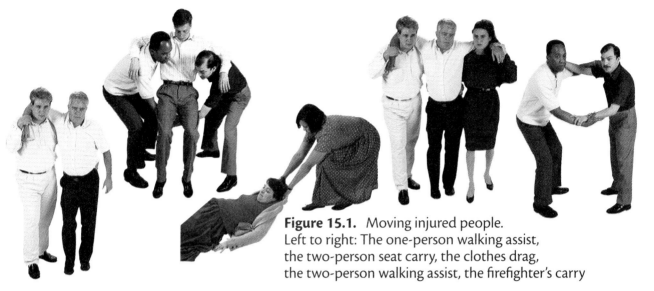

Figure 15.1. Moving injured people. Left to right: The one-person walking assist, the two-person seat carry, the clothes drag, the two-person walking assist, the firefighter's carry

Frostbite occurs when exposure to severe cold causes body tissue to freeze. Skin tissue will look gray or yellowish and feel numb, cold, and doughy. To treat frostbite, warm the affected area gradually. Bring the person into a warm place and put warm water on the site. Do not rub the area. Cover with clean, dry bandages and call for medical help.

Bites

Animal bites from dogs and other animals can **transmit**, or spread, diseases such as **rabies** to humans. If an animal bite occurs, wash the area immediately with soap and water and apply a dressing. Seek medical care if the animal is not a family pet.

Snakebites are serious if they come from one of the four poisonous snakes. These include the rattlesnake, copperhead, water moccasin, and coral snake. If a snakebite occurs, call for help.

LESSON 2 REVIEW Write the answers to these questions on a separate piece of paper. Use complete sentences.

1) What does ICE mean?
2) What are the three types of burns?
3) Define and give the symptoms of heatstroke.
4) How should you treat an animal bite?
5) How would you know if a person was having heat exhaustion?

Frostbite
A tissue injury causing the tissue to freeze due to overexposure to cold temperatures

Rabies
A disease transmitted to humans through animal bites

Transmit
Spread

Health Tip

Keep a first aid kit well stocked with bandages, antibiotic ointment, a chemical ice pack, latex gloves, scissors, and tweezers.

Lesson 3

First Aid for Bleeding, Shock, and Choking

Shock
The physical reaction to injury in which the circulatory system fails to provide enough blood to the body

Some emergencies can be life threatening if first aid is not provided before professional help arrives. Whether a victim lives may depend on the actions of helpful bystanders, such as yourself.

What Is First Aid for Severe Bleeding?

Severe bleeding must be controlled quickly. A rescuer must do four things: (1) stop the bleeding, (2) protect the wound from infection, (3) treat the person for shock, and (4) get EMS help quickly.

If you are providing first aid for severe bleeding, you must protect yourself from possible infections. Use latex gloves if they are available. Cover the bleeding area with a clean cloth. Apply direct, even pressure with the palm of your hand. If blood soaks through the cloth, leave it in place and add more cloth or bandages. Try to elevate the wound above the heart to decrease the flow of blood to that area.

What Is First Aid for Shock?

Severe bleeding, heart attack, and other serious injuries and illnesses can cause **shock**. Shock is a physical reaction to severe injury or illness. Shock can lead to death if not treated, even if the injury is not life threatening. Therefore, always treat for shock immediately, especially if you see the signs of shock. These signs may include a rapid or slow pulse, fast or slow breathing, pale skin, thirst, or confusion.

If a person is in shock, have someone call for EMS help immediately. Then use these first aid steps until help arrives:

1. Keep the person lying down. Elevate the legs higher than the heart to keep blood flowing to the heart. Do not elevate the legs, however, if you suspect there is a spine or neck injury.

Heimlich maneuver
Firm, upward abdominal thrusts that force out foreign objects blocking an airway

2. Maintain a constant body temperature by applying blankets or a coat around the victim.

3. Do not give the person any food or water. This can cause choking.

What Is First Aid for Choking?

Have you ever had food go "down the wrong pipe"? Thousands of people choke to death each year because food or other objects block their airway. Knowing what to do in this emergency can truly be a lifesaver.

A person who is choking may clutch the throat between the thumb and index finger. This is the universal sign for choking. Other signs are gasping, a weak cough, skin turning pale or blue, inability to speak, and possible loss of consciousness. If a choking victim can still speak or cough, let the person try to cough up the object. Never pat the person's back. Doing so will cause the object to go farther down the windpipe and make matters worse.

If the choking victim cannot speak, cough, or breathe, you should do the **Heimlich maneuver**, using the following steps.

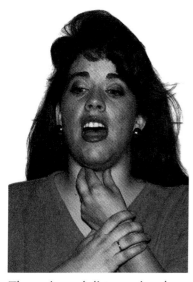

1. Stand behind the choking victim. Wrap your arms around the person's waist.
2. Make a fist with one hand.
3. Place the thumb side of the fist against the middle of the victim's abdomen above the navel and below the ribs.
4. Use the other hand to grab the fist and give quick, upward thrusts into the abdomen.
5. Repeat this action until the object is forced out.

The universal distress signal for choking.

Figure 15.2. First aid for choking

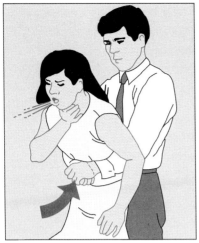
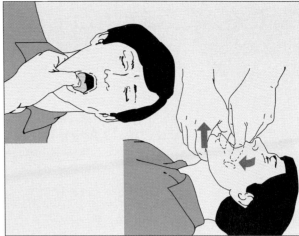

A. The Heimlich maneuver   B. The finger sweep

If the blockage is not removed and the person becomes unconscious, gently lower the victim to the floor. Have someone call 911. Open the victim's mouth and check the throat for the blockage, using the finger sweep. Use the following steps for the finger sweep.

1. Grasp the tongue and lower jaw between your thumb and fingers and lift the jaw.
2. Use your other index finger to sweep the back of the throat in a hooking action, trying to grasp the object.
3. If this doesn't work, begin rescue breathing, which will be described in Lesson 4.

Careers

REGISTERED NURSE

Caring for sick and injured people is just one of the responsibilities of today's nurses. Nurses also teach patients how to care for themselves at home. Some nurses practice law. Others work with state legislatures and government agencies to develop laws that deal with health issues. There are different levels in nursing. To become a licensed practical nurse (LPN) requires one year in nursing school. To become a registered nurse (RN) requires two to four years in nursing school. Other advanced degrees in nursing can also be obtained.

If you are alone and are choking, you can give yourself abdominal thrusts. Hold your hands together as in the Heimlich maneuver and thrust upward against your abdomen. You can also thrust your abdomen against the back of a chair.

LESSON 3 REVIEW Write the answers to these questions on a separate sheet of paper. Use complete sentences.

1) What are four things that must be done when providing first aid for severe bleeding?

2) Why is it important to elevate the legs of someone in shock?

3) Describe the universal sign for choking.

4) Why is it important not to pat a choking person's back while he or she is coughing?

5) If a choking person has lost consciousness and you were able to remove the foreign object, what would you do next?

Lesson 4

First Aid for Heart Attacks and Poisoning

Cardiac arrest
A condition in which the heart has stopped beating and there is no pulse

Cardiopulmonary resuscitation (CPR)
An emergency procedure for cardiac arrest

Irregular
Not normal

As you have read, choking causes someone to stop breathing. Heart attack, drowning, and electrical shock are also causes of stopped breathing. Each of these emergencies requires immediate action.

What Is First Aid for Heart Attacks?

If a blood vessel that supplies blood to the heart becomes blocked, part of the heart becomes damaged. This damage is a heart attack. Signs of a heart attack may include difficulty breathing, **irregular**, or not normal, heartbeat, nausea, or pain in the arm or jaw. Sometimes a person can stop breathing altogether. If a person's heart stops beating, the person is then in **cardiac arrest** and has no pulse. If a person has signs of cardiac arrest, call EMS or 911 immediately.

While waiting for EMS to arrive, you can help the victim by starting **cardiopulmonary resuscitation** (**CPR**). This procedure uses both pressure against the chest and rescue breathing. Pressing the chest helps pump the blood from the heart to the body. Rescue breathing provides oxygen to the lungs so the vessels can transport the blood.

Then and Now

RESUSCI-ANNIE

Cardiopulmonary resuscitation (CPR) was first introduced in the mid 1970s. Prior to this, people would often die while waiting for medical help. What made this profound impact in saving lives was the invention of the Resusci-Annie doll. People could now learn how to do CPR and practice this technique without doing bodily harm as would happen if done on a real person. Resusci-Annie was invented by Asmund S. Laerdal. Laerdal wanted a doll that was life-sized and extremely realistic in appearance so students would be better motivated to learn this lifesaving procedure. Today, the American Red Cross works closely with the American Heart Association to teach people how to save lives using Resusci-Annie dolls.

It is very important to start CPR as soon as possible. Follow these basic steps.

1. Make an open airway.
   - Tap on the person's shoulder or loudly ask, "Are you all right?" to find out if the person is conscious.
   - Tilt the head back and lift the chin to open the airway.
   - Using the finger sweep method, check for any foreign matter that may be blocking the airway.

2. Find out if breathing has stopped.
   - Look, listen, and feel for breathing for three to five seconds.
   - Keep the head tilted back and pinch the nose shut.
   - Take a deep breath, seal your lips around the person's mouth, and give the victim two full breaths. For babies, seal your lips around the nose and mouth and give a gentle puff once every three seconds.
   - Check for a pulse on the side of the neck for five to ten seconds. If you find a pulse, recheck breathing and only proceed if necessary.

What would you do if someone stopped breathing?

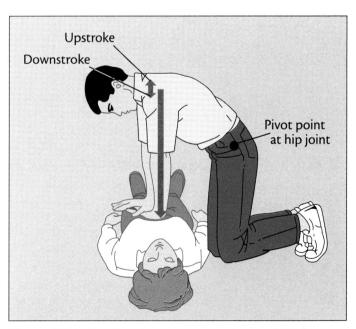

Figure 15.3. Cardiopulmonary resuscitation (CPR)

Compression
The act of pressing down

Oral poisoning
A poison that is eaten or swallowed

3. Find the proper hand position for **compressions**.
   - Locate the notch at the lower end of the breastbone.
   - Place the heel of your other hand on the breastbone next to your fingers.
   - Remove your hand from the notch and put it on top of your other hand.
   - Use only the heels of your hands, keeping your fingers off the chest.

4. Give fifteen compressions.
   - Position your shoulders over your hands.
   - Compress the breastbone one and one-half to two inches.
   - Do fifteen compressions in about ten seconds.
   - Compress down and up smoothly, keeping hand contact with the chest at all times.
   - After every fifteen compressions, give two full breaths.

What Is First Aid for Poisoning?

A poison is a substance that causes injury, illness, or even death when it enters the body. A poison might be swallowed, come in contact with the skin, or be inhaled.

Oral poisoning

Oral poisoning occurs when a harmful substance is swallowed. Household cleaners, radiator fluid, and parts of certain plants, such as morning glories, can be poisonous if eaten. Some signs of oral poisoning are sudden, severe abdominal pain, upset stomach, and vomiting. The victim might become sleepy and lose consciousness.

If a person has swallowed a poison, call the local poison control center immediately. Information about the poison and the victim's weight and age is important to give on the phone. The poison control expert will give you instructions on what to do until emergency medical help arrives. Look on the bottle of liquid you think is the poison to see if it says "Do not induce vomiting." This means the chemical will cause more burning in the throat if vomiting occurs.

Contact poisoning
A poison that comes in contact with the skin

Inhalation poisoning
A poison that is inhaled

Contact poisoning

Contact poisoning occurs when a poison comes in contact with the skin. Poisons may be absorbed from plants such as poison ivy, household cleaning products, or lawn and farm chemicals. Signs of this kind of poison are severe rash, swelling, blisters, itching, and burning. Remove any clothes that have contacted the poison. Wash the skin with soap and large amounts of water. You can apply ice for swelling or calamine lotion to reduce itching.

Inhalation poisoning

Inhalation poisoning occurs when someone breathes in, or inhales, harmful gases. Signs of this type of poisoning are headache, dizziness, or loss of consciousness. A person with inhalation poisoning should be moved into fresh air at once. If the victim is unconscious, check for breathing and a pulse. If these are not present, begin CPR immediately.

Remember that this chapter is only your first step toward becoming skilled in giving emergency medical care. Look into taking a first aid class to learn first aid thoroughly.

LESSON 4 REVIEW Write the answers to these questions on a separate sheet of paper. Use complete sentences.

1) What is a heart attack?
2) What are four basic steps of CPR?
3) Describe three kinds of poisoning.
4) What is the first thing to do if someone has swallowed a poison?
5) If a victim of inhalation poisoning is unconscious, what is the first thing to do?

Chapter Summary

- Staying calm in an emergency helps you to think clearly and to perform first aid properly.

- Dialing 911 or 0 for the operator will put you in touch with the local Emergency Medical Service (EMS).

- Good Samaritan Laws protect emergency caregivers. Universal Precautions protect people from contracting infectious diseases while performing first aid.

- There are three types of burns. First-degree burns damage only the outer layer of the skin. Second-degree burns affect the outer and underneath layers of the skin. Third-degree burns require immediate professional attention.

- Overexposure to extreme heat or cold causes trauma to the body. Quick emergency treatment is needed.

- The signs of shock may include a rapid or slow pulse, fast or slow breathing, pale skin, thirst, or confusion.

- The Heimlich maneuver is the quick upward thrust under the ribcage that dislodges a foreign object blocking the windpipe.

- A person is in cardiac arrest if his or her heart has stopped beating and there is no breathing.

- Cardiopulmonary resuscitation (CPR) combines rescue breathing and pressure against the chest to maintain oxygen flow to vital organs until emergency medical help arrives.

- Poisoning occurs when a harmful substance is swallowed, comes in contact with the skin, or is inhaled.

Chapter 15 Review

Comprehension: Identifying Facts

On a separate sheet of paper, write the correct word or words from the Word Bank to complete each sentence.

WORD BANK

- cardiac arrest
- cardiopulmonary resuscitation (CPR)
- first aid
- fracture
- frostbite
- Good Samaritan Laws
- heatstroke
- Heimlich maneuver
- hypothermia
- poison
- shock
- splint
- sprain
- Universal Precautions

1) A _____ is the tearing or loosening of ligaments or tendons.
2) _____ is the emergency care given to a sick or an injured person.
3) A bone that is broken is called a _____.
4) The _____ protect an emergency caregiver.
5) A _____ can be swallowed, come in contact with the skin, or be inhaled.
6) _____ is the freezing of tissue due to excessive exposure to cold temperatures.
7) A person whose heart stops beating is in _____.
8) A _____ is used to keep a broken limb in place.
9) Overexposure to severe cold temperatures causes _____.

10) Using latex gloves and masks are examples of how _____ can protect a person from contracting infectious diseases.

11) The inability to sweat is a main sign of _____.

12) While waiting for emergency medical help to arrive, perform _____ for anyone whose heart stops beating.

13) To save a person from choking to death, the _____ should be performed.

14) Covering a person with blankets will keep a trauma victim from going into _____.

Comprehension: Understanding Main Ideas

Write the answers to these questions on a separate sheet of paper. Use complete sentences.

15) What are three basic guidelines for first aid?

16) Name four types of problems that may require first aid. Describe the treatment for each type of problem.

17) What is the Heimlich maneuver?

18) How would you start CPR on someone who is having a heart attack?

Critical Thinking: Write Your Opinion

19) What would you do if you were baby-sitting and the child was bitten by a neighbor's dog?

20) What advice would you give someone who is eating large pieces of meat while talking?

Test Taking Tip After you have taken a test, go back and reread the questions and your answers. Ask yourself, "Do my answers show that I understood the question?"

Chapter 16

Preventing Violence

It is becoming harder to avoid violence. Violence is shown on TV nearly every day. You can read about it in newspapers and magazines. The words in some songs and the images in some pictures are violent. It seems that violence is being talked about or shown everywhere. So why do we need to study violence in health class? Violence is a growing threat to the health of young people. Violence can affect young people even more strongly than it does adults. Before we can prevent violence, we need to understand it.

In this chapter, you will learn about different types of violence. You will look at the causes of violence. You will read about victims and the high costs of violent acts. You will learn how people can respect each other and avoid violence. People can avoid violence by identifying problems and solving them in peaceful ways.

Goals for Learning

▶ To define violence and describe its costs
▶ To identify warning signs of conflict
▶ To describe causes of violence
▶ To explain ways to prevent violence and resolve conflicts

Lesson 1

Defining Violence

Media
Sources of information and entertainment, such as newspapers and TV

Violence
Actions or words that hurt people or things they care about

What Is Violence?

Violence is hard to define. Many people feel they "know it when they see it" but cannot put it into words. Violence can be described as any actions or words that hurt people or things that people care about. Violence makes people feel bad about themselves. They may not even feel safe where they live or work.

What Is Violent Behavior?

Violent behavior is acting in a way that hurts others. It can include pushing, hitting, or name-calling. Another violent behavior is damaging someone's property or using a weapon. Violence never solves problems. Instead, it creates new problems that are even worse. Suppose you bumped into someone accidentally. What if you said something that hurt someone's feelings, but you didn't mean to? Is this violence?

Usually when we talk about violent behavior, we mean actions that were intended to do harm. However, we might describe an automobile crash as violent, even if it wasn't anyone's fault. When an accident happens, it is a good idea to think about whether it could have been avoided. Could you have been more careful? Could you have chosen your words more thoughtfully before you spoke? One way to avoid violent behavior is to think about what the results of any of your actions might be.

What Are Some Forms of Violence?

Media Violence

One kind of violence is the things we see and hear in the **media**. Newspapers, TV, video games, and movies contain violence. Some of the violence reported is about true things that happen. Some of it is made-up violence that is meant to make a story seem more real. Many people believe that violence on TV or in movies can make children act violently. When children see violence in the media, they may think that it is acceptable. They may then imitate the violent actions they have seen.

Family Violence

Violence in the family is a problem for many people. Parents may become violent toward one another or toward a child. This may be learned from the parents. Suppose a person was hurt or abused as a child. Then the violent behavior may be repeated with that person's own children.

Random Violence

Sometimes violence is not directed at a specific person. Someone can damage property by spraying it with paint or throwing a rock at it. The person may not have intended to hurt the property owner, but he or she still acted violently. This is called random violence. People can be injured by random violence. When weapons are used on the street or in other public places, people can be hurt or killed.

There may be times when you can pressure friends not to fight. Let them know you support peaceful solutions to disagreements.

Conflict
A disagreement or difference of opinion

Internal conflict
A situation in which a person doesn't know what to do

Health Tip

Try to find a compromise to solve conflicts peacefully.

How Is Conflict Different From Violence?

Violence is often the result of conflict, but it doesn't have to be. **Conflict** is a disagreement between people. These people may have conflicting ideas about something or want to do something a different way. Conflict is everywhere. Disagreements are normal. Turning to violence is not normal.

Some conflicts are bigger than others. You may disagree with a family member about who should do the chores. Your parents may have different ideas about where to go on vacation. Conflicts can be ongoing or brief. Two schools may have a sports rivalry going for years. Friends may have a different plan about which bus to take to the store. You can even have a conflict within yourself. An **internal conflict** is having two different ideas and not being sure which is best.

Some conflicts have easy solutions. Others take time and care to work out. Government leaders face major conflicts when they decide how to rule a nation. A person is faced with a major conflict if challenged to a fight. The way we respond to conflict determines whether violence will occur.

Who Does Violence Affect?

Do you think that most violent acts happen among strangers or friends? Unfortunately, violence can happen in many situations. About half of the murders in the United States happen among people who know each other.

Since 1992, an average of 4.3 million Americans have been victims of violent crime. These crimes include robbery, assault, and other attacks. Many of the crime victims were between the ages of 12 and 24. Why do you think so many violent acts happen to young people?

What Are the Costs of Violence?

When a person is robbed, there is a loss of money or the cost of property. However, the victim may also suffer an emotional loss. He or she may feel unsafe walking down the street. Victims may feel mental or emotional scars. They may have strong feelings of fear, anger, or sadness. Victims may need to see a professional to treat these mental injuries.

Cycle
A repetition

Violence can cause physical harm. Violent acts can cause permanent injuries. Victims may need medical care to heal. Friends and relatives of the victim are affected. They may need to take responsibility for the care of the victim.

The people who commit violent acts may also feel emotional costs for their actions. They may have guilt over what they have done. This might lead them to feel even worse about a situation. If these people do not deal with these feelings, they may resort to violence again.

A person could be arrested for violent behavior. The person may have to go to court and serve time in prison. Once a person has a criminal record, it may be hard to make positive changes in life. A violent past can lead to a **cycle**, or repetition, of violence.

In the end, everyone pays for crime and violence. The costs for police, courts, and prisons add up. When violence occurs, we are all victims.

LESSON 1 REVIEW Write the answers to these questions on a separate sheet of paper. Use complete sentences.

Why might a victim feel anger? What could that person do to feel better?

1) How might the media contribute to violence?
2) What kind of violence is not directed at a specific person?
3) Do conflicts always lead to violence?
4) What are some of the costs of violence?
5) How might seeing violence on TV affect young people's behavior?

Lesson 2

Causes of Violence

When Does Conflict Lead to Violence?

You cannot always avoid conflict. We all face difficult situations sometimes. When problems occur, we sometimes do not see how they can be resolved. At times people do not even realize they are in a conflict. A disagreement can quickly turn into an argument with shouting and angry feelings. If people do not stop to think about their actions and words, anger could turn into violence.

Understanding a situation helps you to respond in the right way. Then you can work to find a solution. How do you know when a disagreement is turning into a conflict that needs to be resolved? Below are some warning signs of conflict. If you notice one or more of these signs, work with others to find out what the problem is.

WARNING SIGNS OF CONFLICT

- Shouting
- Name-calling
- Insults
- Mean looks
- Making fun of others
- Making threats
- Pushing or hitting

What Causes Violence?

The media often show or describe violent actions or words. This is in part just a reflection of our society. However, when young people see violence, they may think that it is a good way to solve problems.

Sometimes parents can set a bad example for their children. Parents may feel stress from work and other responsibilities. Increased stress may make parents respond to problems through violent actions or words. The child may take this as an example of how to deal with conflicts. Many criminals come from violent homes. They were not able to break the cycle of violence from their childhood.

Health Tip

If you are feeling angry, try going for a run or punching a pillow.

The examples that parents set influence their children's behavior patterns.

Many people buy guns, which they keep in the home. They may not intend to use the gun. Usually gun owners feel that guns offer protection and peace of mind. Guns, however, are also dangerous to young people. If a child finds a gun, he or she may think that it is a toy. A gun in the home increases the chance of accidental death.

Drugs and alcohol can make people act in dangerous ways. People who use drugs or alcohol commit half of all violent crimes. Drugs can also prevent a person from using good judgment. People using drugs may do things that they would never do if they did not use drugs.

Sometimes people act violently to show their loyalty to a group. Members of a gang often do things that they might not do on their own. This is giving in to peer pressure. They may think they are stronger or more powerful when they are with the gang. However, their actions really make them more likely to get hurt or killed.

Prejudice and hatred can lead to violence against a particular group. Prejudice is a person's opinion based on another's race, gender, culture, or religion. A prejudiced person doesn't really know the person in that group. People who commit violence based on prejudice want to prove that they are better. All they really prove is that they are hateful.

Prejudice
An opinion based on a person's religion, race, gender, or culture

Healthy Subjects: Social Studies

VIOLENCE AND GENOCIDE

During World War II, German Nazis arrested many European Jews and placed them in concentration camps. They were imprisoned for no other reason than their religious background. The Nazis also imprisoned other minority groups. The Nazis felt that their race was superior. Living conditions in these camps were poor. People were killed or tortured. Those who survived would be forever scarred by the memory of the camps. This practice of mass killing or torture based on race, culture, or religion is called genocide. After the war, many Nazi soldiers were tried in court for crimes against humanity. Today, the United Nations and other worldwide organizations guard against genocide. Still, genocide has occurred in areas where people of different backgrounds are competing for power. All people of the world must work to make sure that the horrors of Nazi concentration camps are never repeated because of prejudice and ignorance.

People may turn to violence when they feel they have been insulted or hurt. They are trying to get revenge for the pain or sadness they feel. Figure 16.1 shows how revenge only leads to greater violence. A misunderstanding can lead to violent actions when people take revenge instead of sorting out their problems.

What Kinds of Crimes Affect Young People?

Young people are just as vulnerable to crime and violence as anyone. In fact, nearly half of all crime victims in the United States are between the ages of 12 and 24. Some crimes affect young children more than others, such as child abuse and kidnapping.

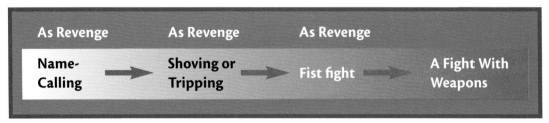

Figure 16.1. Pattern of an insult turning into violence

Often, child abuse comes from within the family. Like other types of violence, child abuse may be part of a cycle of violence. Victims of abuse may be more likely to abuse their own children. Victims of child abuse should get help from an adult, such as a teacher or school nurse. Kidnapping or assault by adult strangers is most likely to happen to children younger than 13 years old. However, it can happen to anyone. Here are some tips for staying away from dangerous strangers:

- Walk only in familiar areas.
- Stay with a group whenever possible.
- Never accept any offerings from strangers. Never go with a stranger anywhere.
- Avoid dark areas such as alleys.
- If you are being followed, walk quickly to a place where there are other people.
- Report any suspicious activity to an adult you can trust, such as a police officer.

LESSON 2 REVIEW Write the answers to these questions on a separate sheet of paper. Use complete sentences.

1) Are gun owners safe from violent crimes?
2) How might name-calling lead to greater violence?
3) Do drugs tend to make people more violent or less violent?
4) Name two kinds of violent crimes that are more likely to affect young people.
5) Why would a person commit violence because of prejudice?

Lesson 3

Preventing Violence

How Can Anger Be Managed?

Everybody gets angry sometimes. Someone may say something that hurts your feelings or do something that you think is unfair. When you feel your body tensing up or your heart beating faster, you probably need to be careful not to get too mad. You might blurt something out that you don't really mean. That could make the situation worse.

Try taking a time-out to relax. Take a deep breath and count to ten. Think about what you want to say and come up with some nonviolent solutions to the problem. If the problem doesn't seem so important after you're relaxed, then you probably shouldn't say anything.

When Should Conflict Be Avoided?

Not all conflicts can be avoided. Some conflicts can be healthy. However, there are some situations where it is best to walk away. It is always a good idea to look at a problem from the other person's perspective. Ask yourself how you would feel if you just forgot about the problem. It may be best to just let it go. You may get along with the other person better the next time.

STOPPING VIOLENCE IN THE SKY

Early airports were much simpler than modern airports. There were fewer flights. There were also fewer concerns about violence in the air. In the 1960s and 1970s, however, terrorist activities placed passengers in danger.

In 1973, new security measures were begun. Under these rules, all baggage was X-rayed. Passengers passed through metal detectors. As a result, less terrorism occurred. By the late 1980s, some airports did not enforce these rules. New terrorism occurred in the 1990s. Again, airports became more careful in examining both passengers and luggage. Such checks keep weapons or bombs from being carried onto planes. The government and airlines do their best to provide passenger safety.

Mediator
A person who helps two sides solve a problem reasonably

What Are Some Methods for Resolving Conflict?

When you decide that it is time to resolve a conflict, you can do some things to make it easier. Try to talk to the person alone or with a **mediator**. A mediator helps people work out a problem reasonably. A mediator does not favor one side or the other.

Listen to what the other person has to say. Do not interrupt, act bored, or daydream. Keep an open mind and consider the person's words. When you explain your feelings, try to remain calm. Keep your emotions under control. Focus on your feelings instead of the other person's actions. If you start to feel upset, ask for a time-out to collect yourself.

How Can You Help Others Avoid Fights?

If you see people fighting, don't encourage it by watching or cheering. That could encourage one of the people to act violently. Let people know that you respect them for being able to walk away from a fight. Give respect to someone who is strong enough to apologize.

STEPS FOR RESOLVING CONFLICT

1. Stay cool. Take a deep breath or go for a walk. Resist acting out in anger.

2. Identify the problem. Take turns listening while the other person talks. Be honest. Avoid insults. Do not interrupt. Ask questions to discover the reasons for the other person's feelings.

3. Use "I" messages to express your feelings. Don't make "you" statements or accusations. That makes the other person defensive.

4. Brainstorm all possible safe solutions. Write down as many solutions as possible. Include even silly solutions. Sometimes they work. For each solution, ask if it is fair, respectful, and true to your sense of right and wrong.

5. Agree on a solution that is fair to everyone. Remember, not every conflict must have a winner and a loser.

6. If you can't resolve the conflict together, perhaps you can seek a mediator.

Peer mediation helps to find solutions to conflicts.

Neutral
Not favoring either side

Sometimes you can take a more active role in preventing violence. You may be able to suggest a peaceful solution to the problem. If people are having trouble working out a solution, offer to be a mediator. Remember, though, that as a mediator you must be **neutral**, not favoring either side. Here are some methods for mediation:

- Let both people know that you are not taking sides.
- Set some rules, such as using calm, reasonable speech.
- Let each person state his or her feelings without interruption.
- Allow each side to ask reasonable questions.
- Encourage them to identify several peaceful solutions.
- Ask them to discuss each solution and agree on one idea.
- Work at a compromise; don't let them give up without really trying.

PHYSICIAN'S ASSISTANT

If you are interested in helping people with health care, you might consider becoming a physician's assistant (PA). A physician's assistant works with a doctor to complete some of a doctor's routine work. PAs get medical information from patients, perform physical examinations, and order laboratory tests. In emergencies, PAs treat cuts, scrapes, bruises, and burns. They know how to set broken bones. Sometimes PAs decide what the results of lab tests mean. They may make diagnoses or prescribe treatments. PAs may handle tasks before or after surgery. Physician's assistants need experience in health care. They also must attend college and complete a recognized degree program.

How Do You Deal With Serious Conflicts?

If you are faced with a serious conflict, don't try to resolve it on your own. Ask a parent, peer mediator, teacher, or counselor for help.

What would you do if you found out that a friend or classmate planned to hurt someone? If you tell someone, it may feel like you are betraying the person. But it is always more important to stop violence than it is to keep a secret. Your friend or classmate will be better off, too, if he or she is prevented from doing something harmful.

Writing About Health

Make a list of some of the conflicts you have faced this year. Describe how you dealt with these conflicts. Are the problems solved now or did they become bigger problems?

LESSON 3 REVIEW Write the answers to these questions on a separate sheet of paper. Use complete sentences.

1) Why should you take a time-out when you feel angry?
2) When is it a good idea to avoid conflict?
3) Why should you avoid "you" statements?
4) Why does a mediator need to be neutral?
5) What are the steps for resolving conflict?

Chapter Summary

- Violence doesn't solve problems. It usually creates even worse problems.

- Conflicts and disagreements are normal. They may be big, small, or ongoing. It is not normal or healthy for a conflict to turn violent.

- Some of the costs of violent crimes are permanent injuries, mental scarring, lost peace of mind, and money.

- It is helpful to identify conflict and potential problems. This is the first step to resolving potential problems and avoiding violence.

- Some potential causes of violence are the media, family problems, guns, drugs, and gangs. Prejudice and the desire for revenge can also motivate violence.

- Young children can be affected by crimes such as robbery and assault. They are also vulnerable to child abuse and kidnapping.

- Everyone feels anger sometimes. People can use a time-out to manage their anger.

- Sometimes it is best to walk away from conflicts.

- It is best to talk alone with a person when resolving conflict. Each side should remain calm, avoid insults, and focus on feelings rather than actions.

- You can prevent others from fighting by not watching or encouraging them. You can also suggest that they work out a peaceful solution.

- Some problems are too difficult for you to work out on your own. It is okay to ask for help from an adult or peer mediator.

- It is more important to stop violence than it is to keep a secret. If you know that a friend may hurt someone, tell a trusted adult and try to stop the violence.

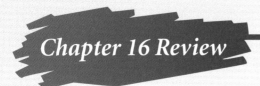
Chapter 16 Review

Comprehension: Identifying Facts

On a separate sheet of paper, write the correct word or words from the Word Bank to complete each sentence.

WORD BANK	
apology	mediator
conflict	mental
encourage	ongoing
imitate	prejudice
increase	revenge
intended	secret
loyalty	shouting

1) Violence is any actions or words that are _____ to do harm.

2) Children may _____ what they see on TV.

3) _____ is a normal disagreement or difference.

4) Some conflicts are _____, lasting for years.

5) A crime victim may have both physical and _____ injuries.

6) _____ is a warning sign of conflict.

7) Guns _____ the chance of violence in the home.

8) People may act violently to show _____ to a gang.

9) _____ usually leads to increased violence.

10) An _____ often helps both sides feel better.

11) _____ and hatred can lead to violence against a particular group.

12) Don't _____ a fight by watching it.

13) A _____ should not choose sides.

14) It is more important to prevent violence than it is to keep a _____.

Comprehension: Understanding Main Ideas

Write the answers to these questions on a separate sheet of paper. Use complete sentences.

15) Give some examples of the media.

16) What is the cycle of violence?

17) What is the term used for violence aimed at a group of people with a particular cultural or religious background?

18) What does a mediator need to be?

Critical Thinking: Write Your Opinion

19) Why might using alcohol make a person less likely to compromise and resolve a conflict?

20) As a mediator, how might you solve a problem if one person is prejudiced?

Test Taking Tip Review your corrected tests. You can learn from previous mistakes.

Deciding for Yourself

Does Gun Control Prevent Violence?

Many violent crimes involve the use of guns. Sometimes the guns were purchased for legal reasons and were not locked up properly. Some of the guns are stolen or bought illegally.

Some people believe that more gun laws will reduce violence. They work to pass laws such as the Brady Handgun Violence Protection Act. This law requires a five-day waiting period before a person can buy a gun. Other such laws may prohibit people from carrying any concealed weapon. Here are some other arguments in support of gun control.

- Gun laws can help stop convicted criminals and drug abusers from buying guns.
- Without gun control, our schools and neighborhoods will become more violent.
- Americans no longer need guns for self-protection. Today we have police departments to protect us.

Other people believe that gun laws are a bad idea. They think everyone should have the right to buy a gun. Here are some arguments against gun control.

- Waiting periods and background checks won't stop someone who wants a gun.
- Gun laws won't decrease violence and crime.
- Every American has the right to self-protection.
- Americans should have the right to use guns for hunting and other recreational purposes.

Questions

1) Which statements supporting gun control do you believe? Do you have other reasons to give in support of gun control?
2) Which statements against gun control do you believe? Do you have other reasons to give against gun control?
3) Do you think we need more laws controlling the buying and using of guns? Support your answer with reasons.

Unit Summary

- To reduce chances of car injuries, don't drive with someone who has been drinking alcohol or using drugs. Serious injuries from a car accident can be reduced by wearing seat belts.

- Wearing proper safety equipment prevents injuries when doing exercise or sports. Warm up and cool down before and after exercising. Drink plenty of water.

- Hand guns and rifles should be in locked cabinets. The ammunition should be stored in a separate, and also locked, location.

- Wiring, appliances, and smoke detectors should be checked twice a year. Checking the house for fire hazards reduces the number of fires. Knowing what to do if a fire breaks out reduces your risk of injury.

- When baby-sitting, keep your full attention on the children. Know what to do in an emergency.

- An emergency kit should have flashlights and batteries, candles and matches, a battery-operated radio, and a first aid kit.

- Plan ahead for what you will do if your area has an earthquake, tornado, hurricane, flood, or severe thunderstorm.

- To get emergency help, dial 911 or 0. Stay calm and talk slowly. Never hang up until you are told to do so.

- The Good Samaritan Laws protect people who help victims in an emergency.

- First-aid providers should use Universal Precautions when possible. These include wearing masks and latex gloves.

- Heat exhaustion, heatstroke, hypothermia, and frostbite occur in very hot or cold temperatures.

- Severe bleeding, shock, or a victim not breathing require immediate emergency first aid.

- The Heimlich maneuver is used for people who are choking.

- Cardiopulmonary resuscitation (CPR) is emergency first aid for heart attacks.

- It is more important to stop violence than to keep it a secret. If you know a friend who may hurt someone, tell a trusted adult.

Unit 6 Review

Comprehension: Identifying Facts

On a separate sheet of paper, write the correct word or words from the Word Bank to complete each sentence.

WORD BANK		
batteries	firearms	shouting
choking	first aid	smoke detector
conflict	prejudice	temperature
conscious	pressure	violence
emergency	seat belts	

1) Wearing _____ can help prevent injuries in car accidents.

2) A _____ can protect you from injury in a household fire.

3) You should dial 911 in an _____.

4) Keep all _____ in locked cabinets.

5) An emergency kit should include a flashlight, _____, a first aid kit, and a radio.

6) Immediate emergency care given to a sick or an injured person is _____.

7) To find out if someone is _____, tap them on the shoulder or ask loudly if the person is OK.

8) For shock, maintain normal body _____ by using a blanket or a coat.

9) The Heimlich maneuver is used when a person is _____.

10) To stop severe bleeding, apply _____ directly to the wound.

11) Hurting people or things people care about is _____.

12) A disagreement between people is a _____.

13) Warning signs of conflict are threatening gestures or _____.

14) Violence aimed at a particular cultural or religious group may result from _____.

Comprehension: Understanding Main Ideas

Write the answers to these questions on a separate sheet of paper. Use complete sentences.

15) Describe at least two actions that can prevent poisoning.
16) Describe what you should do in a tornado or during an earthquake.
17) What is the most important thing to do in an emergency?
18) What is a poison control center and when should you call it?

Critical Thinking: Write Your Opinion

19) Describe two ways you could make your own home safer.
20) What are some ways to avoid becoming the victim of a violent crime?

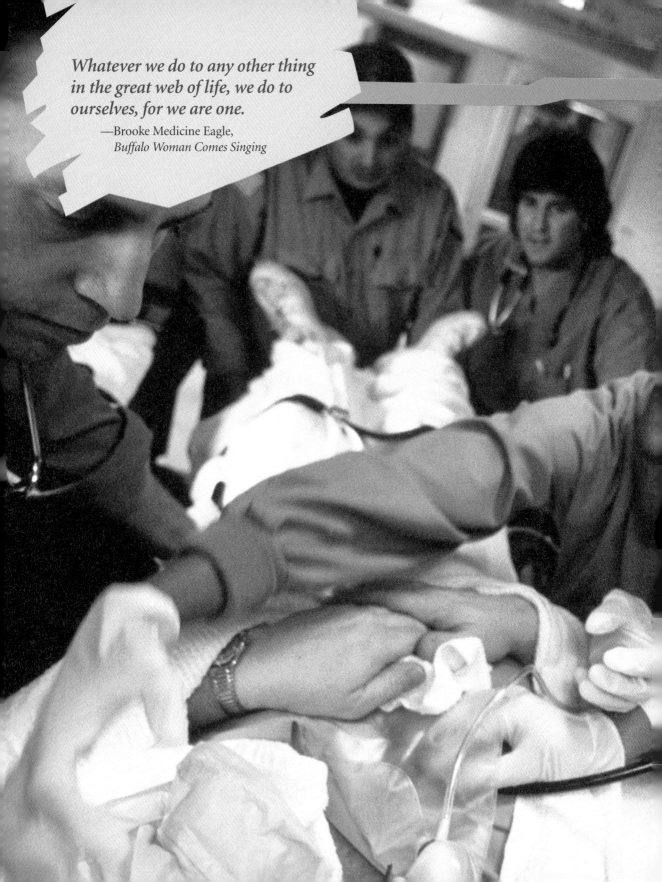

> *Whatever we do to any other thing in the great web of life, we do to ourselves, for we are one.*
> —Brooke Medicine Eagle,
> *Buffalo Woman Comes Singing*

Unit 7

Health and Society

Are the products you buy safe for the environment? Do you help others in your community? Do you take care of yourself to prevent disease? We all have a responsibility to do things that promote public health and safety. We can be wise consumers by buying products that are safe for us and our families. We can promote recycling in our community to protect the environment.

As consumers, we are faced with many decisions that affect our health. We make decisions about choosing health care, doing volunteer work, and preventing air pollution. In this unit, you will learn how to be a wise consumer, promote public health, and protect our global environment.

▶ Chapter 17 Consumer Health
▶ Chapter 18 Public Health
▶ Chapter 19 Environmental Health

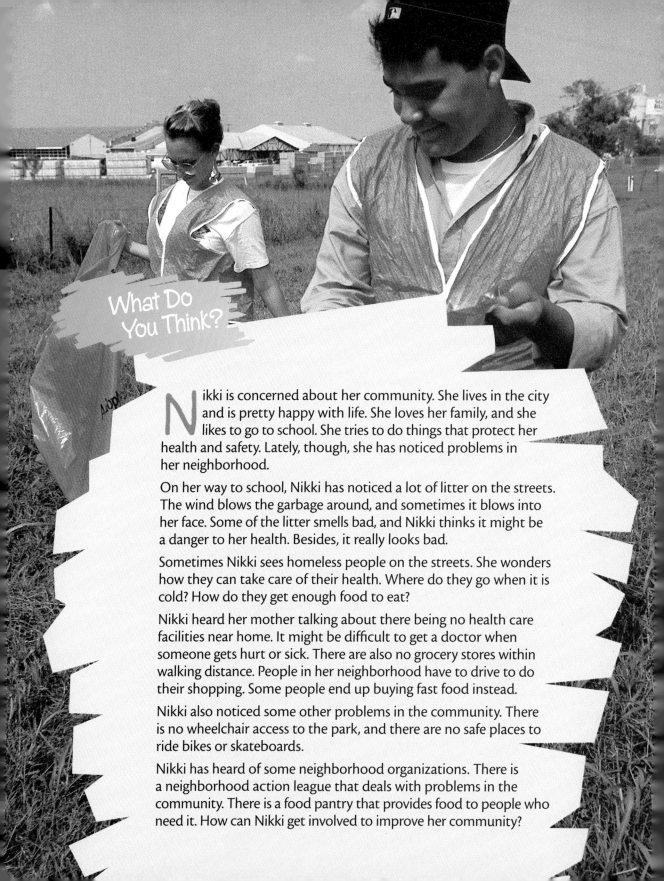

What Do You Think?

Nikki is concerned about her community. She lives in the city and is pretty happy with life. She loves her family, and she likes to go to school. She tries to do things that protect her health and safety. Lately, though, she has noticed problems in her neighborhood.

On her way to school, Nikki has noticed a lot of litter on the streets. The wind blows the garbage around, and sometimes it blows into her face. Some of the litter smells bad, and Nikki thinks it might be a danger to her health. Besides, it really looks bad.

Sometimes Nikki sees homeless people on the streets. She wonders how they can take care of their health. Where do they go when it is cold? How do they get enough food to eat?

Nikki heard her mother talking about there being no health care facilities near home. It might be difficult to get a doctor when someone gets hurt or sick. There are also no grocery stores within walking distance. People in her neighborhood have to drive to do their shopping. Some people end up buying fast food instead.

Nikki also noticed some other problems in the community. There is no wheelchair access to the park, and there are no safe places to ride bikes or skateboards.

Nikki has heard of some neighborhood organizations. There is a neighborhood action league that deals with problems in the community. There is a food pantry that provides food to people who need it. How can Nikki get involved to improve her community?

Chapter 17

Consumer Health

Health care products and services are available everywhere. Most people buy them every day. You buy health care when you use a product or a service that directly affects your physical and emotional well-being. You buy health care whenever you see a doctor, dentist, or nurse. Becoming a wise health buyer takes effort. It means learning how to judge the products and services you use. It means knowing when and where to seek health care and ways to pay for it.

In this chapter, you will learn what wise health consumers know and what a consumer's rights are. You will learn what influences people's health choices, how to judge products and services, and how to clear up consumer problems. You will learn when to use self-care and when and where to seek professional health care. You will learn how health care expenses are paid.

Goals for Learning

▶ To identify sources for health care
▶ To learn when self-care or professional treatment is best
▶ To explain ways to pay for health care
▶ To explain the advantages of being a wise health consumer
▶ To describe ways to judge products and services
▶ To describe how advertising influences health care consumers

Lesson 1

Health Care Information

> **Specialist**
> *A doctor who works only on certain types of medical problems*

As a person who buys health care, you need to be aware of what is available to you. You also need to know how you can get the health care you need.

What Is Health Care?

Health care includes many different activities, including preventive medicine and regular doctor checkups. Sometimes special measures need to be taken to preserve your body's health. You may need a medical procedure, an operation, or emergency medical care to treat an illness or a disease.

However, health care is also something that you can do every day. When you eat a balanced diet and exercise regularly, you are practicing preventive medicine. Using grooming products to keep your body clean and healthy is another form of health care. If your doctor has prescribed medication or recommended vitamin supplements, then taking these is also a part of your health care.

Who Can Help You With Health Care?

You probably know about medical professionals such as doctors or nurses. You may have been to a medical **specialist**, such as an orthopedic surgeon or a pediatrician. However, there are many other people in the health care industry. Pharmacists at the drugstore can help you pick out over-the-counter medicine, or things you can buy at the store. They will also help you understand how to take any medication your doctor has prescribed. Employees at an insurance company can help their customers understand what health care is available to them and how they can pay for it. There are also many workers at companies that make health care products. All these people are a part of our health care system.

Health Tip

Make regular appointments with a doctor and dentist twice a year, and an eye doctor once a year.

346 *Chapter 17 Consumer Health*

What Health Care Products Are Available?

Grooming products, over-the-counter medicine, and prescription medication are some health care products that you may already be familiar with. Health care centers may have more specialized types of products. Most hospitals and clinics carry equipment, machines, and computers that are not available elsewhere. From medical syringes to heart monitors, many products can help diagnose and treat illness or disease.

How Is Medical Care Given?

Medical care can be given from a doctor's office, clinic, or hospital. For routine procedures, you probably would want to go to an office that is near your home. If your medical care is managed through an insurance company, you may need to see a doctor who is in the company's plan.

Make regular appointments with your doctor.

LESSON 1 REVIEW Write the answers to these questions on a separate sheet of paper. Use complete sentences.

1) How is eating a balanced diet a part of health care?
2) Name three types of health care professionals besides doctors and nurses.
3) What are some health care products you can buy?
4) Where would you go for a routine medical checkup?
5) How does a pharmacist assist a doctor in administering health care?

Consumer Health   Chapter 17   347

Lesson 2

Seeking Health Care

Prevention
Appropriate, ongoing self-care

Wise health consumers know how to take care of themselves. They also know when it is best to see a doctor.

When Is Self-Care Appropriate?

One way to take care of yourself is to use **prevention**, or appropriate, ongoing self-care. Good nutrition, regular exercise, and safety practices help to prevent illness or injuries. Not smoking and not being around smokers are also preventive measures. Prevention also means not using tobacco, alcohol, or any other drugs unless they have been prescribed by a health professional. Keeping a healthy body weight that's right for your height and age can help to prevent heart and other problems.

It is a good idea to use self-care for colds and other minor illnesses. For example, sound self-care for a cold is to drink plenty of liquids and get extra rest. If a minor illness or injury does not get better within a few days, you should consult a doctor.

When Should People Seek Professional Help?

Sometimes it is best not to treat your own health problem. As a rule, consult a health professional whenever you notice something unusual about your body, behavior, or thoughts. Here are some examples of situations for which you should seek professional help:

- An injury that is more than a minor cut or bruise or any blow to the head
- Any unusual bleeding, such as blood in the urine or feces
- Any sharp pain, such as in the stomach
- A patch of skin or a mole that has changed shape or color
- A minor problem—such as a cold, flu, cough, or sore throat—that lasts more than a week
- Feelings of sadness that last more than a few days

QUESTIONS TO ASK YOURSELF ABOUT A DOCTOR

- Does the doctor ask about your health problem and history?
- Does the doctor listen carefully and answer all your questions?
- Do you feel comfortable with the doctor?
- Does the doctor give you enough time and attention?
- Does the doctor clearly explain the purpose of tests and procedures?
- Does the doctor clearly explain the tests or diagnosis?
- Does the doctor encourage you to take preventive action?
- Is the doctor willing to refer you elsewhere when necessary?

Nurse practitioner
A registered nurse with special training for providing health care

Primary care physician
A doctor who treats people for routine problems

Who Are Some Health Care Professionals?

There are many choices for professional health care. Most often people see a doctor for health care. A **primary care physician** treats people for routine or usual problems. For example, a primary care physician may do preventive checkups and handle illness or injury such as the flu or a muscle sprain. A primary care physician usually has a focus in family or general practice. If a problem is too difficult for a primary care physician to handle, he or she may refer you to a specialist.

People often see a **nurse practitioner** for minor problems. A nurse practitioner is a registered nurse with special training for providing primary health care. The nurse practitioner can do many of the things that a doctor does.

Sometimes doctors refer people to other kinds of health care providers. For example, a doctor might refer someone to a physical therapist or a registered dietitian. A physical therapist helps people to regain their muscle functions. A registered dietitian provides nutritional counseling for people who have special dietary needs. A dietitian might help someone with diabetes.

MEDICAL SPECIALISTS

Medical specialists need special training so that they understand and can treat all the conditions and problems related to their field. Here are the names of some medical specialists and descriptions of what they do.

Allergists diagnose and treat allergies such as hay fever and hives. Dermatologists diagnose and treat skin disease. Obstetricians deal with pregnancy and childbirth. Oncologists treat cancer and its symptoms. Ophthalmologists diagnose and treat eye problems. Orthopedic surgeons diagnose and treat bone and joint problems. Pediatricians provide primary care for babies, children, and adolescents.

Have you been to any medical specialists? Did they use special equipment or a procedure that was new to you?

Health care facility
A place where people go for medical, dental, and other care

Inpatient
Someone receiving health care who stays in a facility overnight or longer

Outpatient
Someone receiving health care without staying overnight

Pediatrics
Child health care

Rehabilitation
Help to recover from surgery, illness, or injury

How Can You Find a Health Care Professional?

One way to find a health care professional for your needs is to ask a trusted person for a recommendation. Family members or friends may know someone who they feel is qualified. Another doctor or nurse could also give you a recommendation. Directories in the public library list physicians, their specialty, and where and when they received their training.

Once you have a list of names, you can decide what kind of qualities are important to you. Do you prefer a male or female health care provider? Do you want to see a young health professional or one with more years of experience? If you feel uncomfortable with a health care provider, you may want to choose a different provider.

What Are Some Health Care Facilities?

A **health care facility** is a place where people go for medical, dental, or other types of care. A dental or medical office is one kind of health care facility. You may go to a medical office for routine checkups and to test and diagnose injuries, diseases, or other problems.

A clinic may provide **outpatient** care for people who do not need to stay overnight. Outpatient clinics offer primary care and same-day surgery. The patient can go home the same day. Most clinics treat all kinds of medical problems. Others have a specific focus, such as **pediatrics**, or child health care.

When receiving **inpatient** care, a person must stay overnight or longer. Sometimes patients need to recover from surgery, an illness, or an injury. They can receive nursing care and possibly **rehabilitation**, or help to get better, while in inpatient care.

A hospital is equipped to provide complete health care services. People can receive both outpatient and inpatient care at a hospital. Hospital emergency rooms offer care twenty-four hours a day. These centers treat people who cannot wait for an appointment at a doctor's office.

Sources for Health Care

Health Care Professionals	Health Care Facilities	Insurance
Dentist Treats routine or usual teeth problems	**Dental or Medical Offices** Provide routine examinations and tests to diagnose and treat injuries, diseases, and problems	**Health Maintenance Organization (HMO)** Provides care within certain limits to enrolled members and their families
Dermatologist Diagnoses and treats skin problems	**Hospital** Provides complete health care services	**Preferred Provider Organization (PPO)** Provides their members more coverage if they choose health care providers in the plan
Doctor Diagnoses and treats routine or usual problems	**Long-Term Care Facility** Provides health care for people who require extended recovery and rehabilitation after surgery	
Nurse Practitioner Provides primary health care		**Medicare** Provides insurance for people age 65 or older
Physical Therapist Helps people regain their muscle functions	**Outpatient Clinic** Provides primary health care for people who do not need to stay overnight	**Medicaid** Provides medical aid for people whose incomes are below an established level
Psychiatrist Diagnoses and treats mental and emotional problems		
Registered Dietitian Provides nutritional counseling		

Hospice
A long-term care facility for people who are dying

Terminally ill
Dying from a disease, an injury, or an illness, sometimes over a long period

Nursing homes and other long-term care facilities offer everyday care for people who cannot care for themselves. They help older adults or people living with a disability. Long-term care facilities can also help people who require extended recovery and rehabilitation after surgery. A **hospice** is a kind of long-term care facility for people who are **terminally ill**, or certain to die from a disease.

LESSON 2 REVIEW Write the answers to these questions on a separate sheet of paper. Use complete sentences.

1) What kind of health care is prevention?
2) When is it wise to seek professional medical help?
3) What does a primary care physician do?
4) What are the two kinds of health care facilities?
5) Why would a primary care physician need to refer you to a medical specialist?

Lesson 3

Paying for Health Care

Deductible
The initial amount a patient must pay before insurance covers health care costs

Health insurance
A plan that pays all or part of medical costs

Out-of-pocket
Straight from a person's income or savings

Premium
An amount of money paid to an insurance company at regular intervals

It can be very expensive to pay for health care. When someone suffers from an injury or illness, the care and equipment needed to make him or her well can be very costly. Some people can pay for their medical expenses **out-of-pocket** from their income or savings. Most people, however, have **health insurance** to cover their medical expenses.

How Does Health Insurance Work?

Health insurance is a system where you pay a **premium**, or a set amount of money to a company at regular intervals. The company then pays for all or part of your medical costs. This may include medicines, surgery, tests, hospital stays, and regular doctors' visits. Health insurance covers large medical expenses that people might otherwise not be able to afford. Three major kinds of health insurance are private, managed care, and government supported.

Private Health Insurance

In private insurance, people pay for a policy themselves or through their employer. An individual may also need to pay a **deductible**, or an initial amount before the insurance pays the costs. For example, if a deductible is $100, you must pay the first $100 of your medical expenses each year.

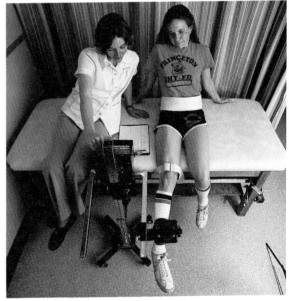

Health insurance covers larger medical expenses that people might otherwise not be able to afford.

352 Chapter 17 Consumer Health

Medicaid
Health insurance for people with low incomes

Medicare
Health insurance for people age 65 or older or who receive Social Security disability

Managed Care

People in a managed care plan usually pay a premium, just as in private insurance. However, they may not have to pay a deductible as long as they see medical professionals who are part of the managed care plan. Plans usually have a number of qualified professionals to choose from.

Government-Supported Health Insurance

The U.S. government provides aid to people who otherwise may not be able to afford health care. **Medicare** provides insurance for people age 65 or older and people on Social Security disability. It pays for hospital and nursing home care. Individuals must pay for some uncovered expenses, such as physician care. **Medicaid** provides similar aid for people with incomes below a certain level.

How Can Individuals Lower Health Costs?

There are many ways to keep medical expenses down. The cheapest form of health care is preventive medicine. Regular doctor visits and maintaining a healthy lifestyle are two ways of avoiding costly medical treatments.

Outpatient care is almost always less expensive than inpatient care, which requires overnight stays. When possible, you can choose procedures that allow you to leave the hospital the same day.

You can also receive health care from low-cost facilities. Neighborhood health clinics, alternative birthing centers, and hospices are some places that provide various health care services at a lower cost than most hospitals.

Health Tip

If your health insurance has a preventive care policy, take advantage by going for regular checkups.

LESSON 3 REVIEW Write the answers to these questions on a separate sheet of paper. Use complete sentences.

1) What is a deductible?
2) What are three kinds of health insurance?
3) Who qualifies for Medicare?
4) Which is cheaper, inpatient care or outpatient care? Why?
5) How does health insurance help people receive health care?

Lesson 4

Being a Wise Consumer

Consumer
A person who buys goods and services

Defective
Not working properly

To stay healthy, you need to be a wise **consumer**. A consumer is someone who buys goods or services. For example, you buy goods such as soap or aspirin. You buy services such as dental or medical care.

What Advantages Does a Wise Consumer Have?

As a wise consumer, you can make good choices that protect and improve your health. This means you avoid buying products that are useless or harmful. You recognize signals that you need health care, and you get help quickly.

Wise consumers can also save money by getting the best product or service for the least money. It will take some time to find the best value, but it pays off in the long run. For example, you can check publications with information about consumer products. They will help you compare features, benefits, and prices of different products.

A wise consumer has increased self-confidence. Speaking up for your rights means taking care of yourself. For example, if a product is **defective**, or doesn't work properly, you can make a complaint to the proper source. Usually you can get your money back or have the product replaced. The satisfaction of researching and finding the right product or service also can boost your self-confidence. Knowing and exercising your rights is part of being a wise consumer.

What Influences Consumers' Choices?

Many things influence consumers' choices. Advertisers use techniques that influence people without their realizing it. For example, you may keep singing a catchy tune that you heard advertise a product. Without thinking about the reason, you may buy that product. It takes careful thought to consider how accurate advertising claims are.

Count
The number of items in a package

Quackery
A medical product or service that is unproved or worthless

Word of mouth
Information about a product that you hear from a friend or family member

Another influence on consumers is the advice or opinions of family and friends. For example, if your friend tells you that he likes a certain brand of deodorant, you may decide to try it. You may also buy a product because your family has always used it. **Word of mouth** can be helpful when you consider a purchase. Keep in mind, however, that what is right for another person may not be right for you.

The price of a product can influence your decision to buy it or not. Well-advertised brand-name products are familiar but usually cost more than less-known brands. However, the advertised brand might appeal to a consumer because it has a recognizable name. If consumers are concerned about the safety and quality of a product, they may feel influenced to buy a brand they are familiar with.

The size of a product can affect its price and its overall value. It is important to compare the price with its weight, volume, or **count**—the number of items in a package—to determine the best value.

Ask your pharmacist to advise about the medication you are taking.

Where a product is sold can affect its price. For example, the cost of a product in a convenience store may be higher than in a drugstore or discount store. The wise consumer will shop around for the best price.

How Does Quackery Influence Consumers?

Quackery is the promotion of medical products or services that are unproved or worthless. Each year Americans spend billions of dollars on products that make false claims of curing diseases and reversing conditions.

Anyone can be fooled by quackery. People who are terminally ill, overweight, or unhappy with aging are targets of quackery. Quackery can fool even cautious people.

Bogus
Nongenuine

Generic
Nonbrand-name

People would like to believe that every disease and condition has a cure and that pain and suffering always can be eased. This is not always the case. Sometimes there is no real cure for a medical condition. Other times, these nongenuine, or **bogus**, products prevent people from seeking proper treatment. At best, people waste their money on these products and services. At worst, people who buy fake treatments or worthless medications delay getting appropriate treatment until the problem is severe. People who do not get proper treatment may become sicker. They could even die.

How Can Products and Services Be Judged?

The best way to decide if a product is worth the cost and is right for you is to study its ingredients. This involves making a habit of reading labels. Federal law requires that manufacturers list ingredients on the label of every food, drug, and cosmetic or grooming product. For example, if you were buying a pain reliever, you might look both at brand-name and **generic**, or nonbrand-name, products. Both have labels listing their ingredients. Food labels usually are easy to read. Labels for cosmetic and grooming products and drugs can be more difficult to read. Reading the label provides important information.

The best source of information about a health care product or treatment is your doctor, pharmacist, or a respected agency. A wise consumer should keep in mind that if something sounds too good to be true, it probably is.

LESSON 4 REVIEW Write the answers to these questions on a separate sheet of paper. Use complete sentences.

1) What can you do if a product is defective?
2) How might advertising influence our consumer decisions?
3) How does quackery hurt consumers?
4) What is the best source of information about health care products?
5) What is the best way to decide if a product is right for you?

Lesson 5

Evaluating Advertisements

Most products are advertised on TV and radio or in newspapers and magazines. Companies spend large amounts of money to encourage people to buy their products. Why do they do it? Sometimes people choose products that they have seen advertised. But is this the best way to decide if a product is good or not? What if it affects your health? A wise consumer should look critically at claims made in advertisements before deciding whether to buy a product.

What Are Some Methods of Advertising a Product?

Advertisers use a number of methods and catchphrases to convince you to buy their products.

New Product

Words like *new* and *improved* appeal to the buyer's desire to be different. Sometimes the product is the same; only the packaging is new.

Old-Fashioned

Advertisers appeal to the ideals of tradition and happy memories. Consumers may buy the product because it reminds them of a relative or something from their past.

Healthy Subjects
Consumer Science

CONSUMER REPORTS

The magazine *Consumer Reports* tests and evaluates thousands of products every year. It tests all kinds of household items and other consumer products, from radios and phones to cars and washing machines. It rates different brands by performance, reliability, and specific features. It also provides a chart that shows the rate of repairs for each brand.

Consumer Reports is a useful resource for rating various products and testing the claims they make in advertising. The magazine does not have any product advertising. This is so that it can stay impartial, or fair. The magazine judges the products only on their quality.

Expensive and Superior
These products are more expensive. This appeals to people who want the best.

Inexpensive and a Bargain
Advertisers try to convince people they are getting the best bargain.

What group of people would a cartoon animal character appeal to?

Recommended by Someone Famous
Advertisements use an athlete or a TV or movie star to encourage people to buy a product. Advertisers hope this gives their product instant name recognition.

Healthy Product
Some products claim "no cholesterol" or "no fat" or make other references to health, such as "all natural."

All of these advertising claims should be examined before a consumer makes a decision about a product. Consumers need to be careful, especially when the product affects their health.

technology

ADVERTISING ON THE INTERNET

The Internet has become very popular. Millions of people log on and browse every day. Because of this, many companies want to get advertisements of their products on Web sites. Some of these advertisements are just words and pictures, like in a magazine. Some have graphic animation that is fun to watch. If you click on some of these advertisements, you will be taken to a new Web page with more information about the product.

Some advertisers are now using information from your computer to figure out what you like. They can put advertisements on the Web browser that are tailored to your interests. Some people think this is an invasion of privacy. Others think it would be useful and convenient to have advertisements for things they like. What do you think?

Careers

MEDICAL RECORDS TECHNICIAN

Can you keep track of lots of details? If so, you might want to think about a career as a medical records technician. Medical records technicians gather and report patient information. This may include symptoms, medical history, examination results, lab test results, and treatments. All records must be properly arranged. Technicians translate disease names and treatments into coding systems. It is important to be organized and keep detailed records. Most medical records technicians work on computers and need excellent typing skills. Technicians may work in hospitals, clinics, health agencies, insurance companies, nursing homes, or other health care facilities. A high school diploma with on-the-job training is required. Sometimes a medical records course at a community college may be required. The job outlook for medical records technicians is excellent.

Why Advertise a Health Care Product?

Competition
Many companies trying to sell similar products

Many over-the-counter products advertise because there is so much **competition**. Competition is many companies trying to sell similar products. Advertisements for pain reliever medications use many numbers and figures to try to convince consumers that their product is the best. They may also claim that their product is "recommended by most doctors."

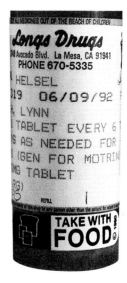

Sometimes hospitals, drug manufacturers, or insurance companies advertise. They may be trying to increase their business, or they may be trying to raise awareness of a new product.

Action for Health

ANALYZING ADVERTISEMENTS

Cut several advertisements from magazines and newspapers, or record or write down the message from a TV or radio show.

What was being sold? What kind of claims does the advertisement make about its product? Is it convincing? Were any health risks portrayed in the advertisements? Was there any depiction of violence or use of drugs, such as alcohol or tobacco? Make your own advertisement for the product. Then think about how you would advertise preventive health care.

Minor
A person under age 18 or 21

What Are the Effects of Advertising Harmful Drugs?

There are many advertisements for harmful drugs such as tobacco and alcohol. Although these products cannot be sold to **minors**, people under age 18 or 21, the advertisements can be seen by anyone. Sometimes these advertisements use cartoon or animal characters to sell their products. Many people have felt that these companies were trying to make their product appealing to children. Recent government regulations put new limits on cigarette advertising, including where the ads can be placed and what they can show.

LESSON 5 REVIEW Write the answers to these questions on a separate sheet of paper. Use complete sentences.

1) Why would a company claim that its product was old-fashioned?
2) How might a famous person help sell a product?
3) Why might a hospital or drug company advertise?
4) Why would a company use a cartoon character to sell its product?
5) What are some problems with advertising?

Lesson 6

Consumer Protection

> **Consumer advocate**
> *A person or group that helps consumers correct problems*

Quackery or other poor medical products or services are a problem for consumers. Quackery may be misleading or confusing. Quackery makes it hard for the consumer to decide what is right and make wise choices. Fortunately, government agencies and **consumer advocates** work to enforce laws and rules that protect consumers.

Consumer advocate groups help consumers find the best way to receive satisfactory answers for their complaints. Consumers can consult these agencies and decide on the best course of action to protect their rights.

What Are a Consumer's Rights?

Consumers have some basic rights that protect them from quackery and other misleading or confusing practices. The U.S. government has established the Consumer Bill of Rights, which gives every consumer these rights:

- *The right to safety*—Consumers have the right to be protected from unhealthy products.
- *The right to be informed*—People can ask for the facts they need to protect them from misleading advertising and to make wise choices.
- *The right to choose*—People have the right to make their own choices.
- *The right to be heard*—When consumers aren't satisfied, they have the right to speak out. This often helps to make laws that protect consumers.
- *The right to redress*—Consumers have the right to get a wrong corrected.
- *The right to consumer education*—People have the right to be educated about the products and services they buy.

Hospital patients are protected by a similar set of rights—the Patient's Bill of Rights. These rights were established by the American Hospital Association. They are shown below. Many individual hospitals, medical centers, and dental and medical offices have established similar lists of rights. You may have been given a copy of such a list or seen one posted.

Patient's Bill of Rights

1. The right to considerate and respectful care
2. The right to complete, current information about your condition in terms you can understand
3. The right to receive all information necessary to give informed consent before any treatment
4. The right to refuse treatment to the extent permitted by law and to be informed of the consequences
5. The right to privacy during examinations and confidentiality concerning your care and records
6. The right to expect a reasonable response when you ask for help
7. The right to be told if your treatment will be part of a research project and to refuse to take part in the project
8. The right to expect good follow-up care
9. The right to an explanation of your bill
10. The right to know hospital rules that apply to your conduct as a patient

How Can a Consumer's Problem Be Corrected?

Sometimes people buy a defective product or have a bad experience with a health care provider. When consumers' rights are violated or broken, they can respond to correct the problem.

The first step is to deal directly with the manufacturer of the product or with the health care provider. Many manufacturers have toll-free numbers. Call and explain your complaint.

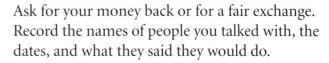

Writing About Health

Think of a time you had a complaint about a product or service. Write what you did about it. Which items in the Consumer Bill of Rights justified your complaint?

Ask for your money back or for a fair exchange. Record the names of people you talked with, the dates, and what they said they would do.

If you aren't satisfied, the second step is to contact one of the agencies listed below. It is best to submit the complaint in writing. Give the details of your problem, including any lack of cooperation from the manufacturer or caregiver. Send the original letter to the appropriate agency.

Sources of Help for the Consumer

For complaints about a product or service, false advertising, or irresponsible behavior:

Better Business Bureau
Check the phone book for
the local bureau.   Web site: www.bbb.org

Also for complaints about false advertising:
Federal Trade Commission Office of Public Affairs
600 Pennsylvania Ave., NW   Web site: www.ttc.gov
Washington, DC 20580   E-mail: webmaster@ftc.gov

For complaints about foods, drugs, or cosmetics:
Food and Drug Administration
5600 Fishers Ln.   Web site: www.fda.gov
Rockville, MD 20857   E-mail: webmail@oc.fda.gov

Also for complaints about food:
Food Safety and Inspection Service
U.S. Department of Agriculture   Web site: www.fsis.usda.gov
1400 Independence Ave., SW   E-mail: fsis.webmaster@usda.gov
Washington, DC 20250

For complaints about dangerous products:
U.S. Consumer Product Safety Commission
Check the phone book   Web site: www.cpsc.gov
for the field office near you.   E-mail: info@cpsc.gov

ELECTRICITY AND HEALING

In 1910, a new product called the White Cross Electric Vibrator Chair promised to help with many problems. These included rheumatism, backaches, and stomach, kidney, hearing, and vision problems. This was quackery, or false medical help, because its claims were not true.

The government began regulating medical devices in 1938. Since then, electricity has been used to treat many illnesses. Small battery-operated devices are placed in the body to help control the heart or other organs. Larger electrical devices produce healing magnetic or radio waves. Treatments with these devices have helped many people with arthritis and Lyme disease.

You should also send a copy to the company or person the complaint is against. Keep a copy for your records. If you don't get a response within six weeks, send a follow-up letter with a copy of the first letter.

What Other Actions Can a Consumer Take?

If other methods of correcting the problem do not work, consumers may consider taking legal action against the manufacturer or health care provider. The government may be able to help make a case against the offender. If the offenses are severe enough, it may even be able to bring criminal charges against the offender.

LESSON 6 REVIEW Write the answers to these questions on a separate sheet of paper. Use complete sentences.

1) What protection do consumers have against quackery?
2) What organization established the Patient's Bill of Rights?
3) What is the first step to correcting a consumer problem?
4) Where can consumers go with a false advertising complaint?
5) How might a consumer group help a person who had been severely injured by an unsafe medical practice?

Chapter Summary

- Health care includes many things, from major surgery to hygiene and healthy living every day.

- The health care industry includes many people in different jobs and many products and services.

- Prevention and taking care of colds and minor illnesses are wise forms of self-care.

- You should consult a health care professional for anything out of the ordinary about your body or behavior.

- Health care professionals include primary care physicians, specialists, nurse practitioners, and providers such as physical therapists.

- You may choose a health care professional based on recommendations from others, medical directories, and your personal preferences.

- Individuals can pay for health care costs out-of-pocket or through private insurance or managed care.

- Government-supported health insurance includes Medicare for people over 65 and Medicaid for people with low incomes.

- Individuals can help lower health care costs by practicing prevention, taking part in their own health care, and choosing low-cost alternatives when possible.

- A wise consumer saves money, protects and improves his or her health, and increases self-confidence.

- Advertisers use a number of methods and catchphrases to convince consumers to buy their product.

- Consumers need to be critical of advertisements and evaluate which claims may be exaggerations and which may be useful information.

- The U.S. government protects consumers with some basic rights. Many health care organizations also set forth rights for patients.

- Wise consumers should read product labels and consult a doctor or other qualified medical professional to determine whether a product is appropriate.

- Consumers who buy a defective product or have a bad experience with a health care provider can follow some steps to correct the problem.

Chapter 17 Review

Comprehension: Identifying Facts

On a separate sheet of paper, write the correct word or words from the Word Bank to complete each sentence.

WORD BANK	
Better Business Bureau	muscle
checkups	pharmacist
competition	prevention
expensive	regular
health professional	rights
hospice	specialist
managed care	value

1) A _____ can help you pick out over-the-counter medicine and understand your prescription.

2) _____ includes keeping a healthy body weight and not using tobacco or alcohol.

3) A person should seek a _____ if the flu or a fever lasts more than a week.

4) You go to a primary care physician for preventive _____.

5) A physical therapist helps people regain their _____ functions.

6) A premium is money paid at _____ intervals.

7) In _____, you may have a limited choice of doctors.

8) Wise consumers find the best _____.

9) Your primary care physician may refer you to a _____ to address a certain illness.

10) An advertiser may say a product is _____ because it is the best.

11) Companies advertise and make claims about their products so that they sell better than the _____.

12) You might go to live in a _____ if you were terminally ill.

13) The American Hospital Association established a list of _____ for patients.

14) If a manufacturer or provider does not respond to a consumer complaint, you could contact the _____.

Comprehension: Understanding Main Ideas

Write the answers to these questions on a separate sheet of paper. Use complete sentences.

15) If you have a low income, how could you pay for health care?

16) If you need to have a medical procedure, what are two ways you could lower costs?

17) Why would a generic brand be cheaper than a name brand?

18) What kind of people might be fooled by quackery?

Critical Thinking: Write Your Opinion

19) Why might a recommendation by word of mouth for a pain reliever be more reliable than an advertisement claim?

20) Why do we need a Consumer Bill of Rights?

Test Taking Tip When taking a matching test, match all the items that you are sure go together. Cross those items out. Then try to match the items that are left.

Chapter 18

Public Health

Public health, or the health of the community, depends on a balance of community resources and personal responsibility. The health of your community is important because it affects the health of every community member.

In this chapter, you will learn what makes a community and how health is a community concern. You will learn about resources that are available and how community members support these resources. You will learn how to help promote public health.

Goals for Learning

▶ To identify what makes up a community
▶ To analyze the importance of health for community members
▶ To identify the health resources available in a community
▶ To explain how resources are funded
▶ To describe ways that community members can make wise decisions to strengthen their community

Lesson 1
Defining Community

Common association
Similar cultural, racial, or religious ties

Common identity
Similar interests or goals

Community
A group of people who live in the same place or have common interests

A **community** is a group of people who live in the same place or share common interests. Strong communities have members who work together for the good of everyone. When problems affect a community, they affect all community members.

How Do You Identify a Community?

Communities come in different sizes. Some small towns may be considered a single community. In larger cities, there may be many communities. Community boundaries may be created by natural formations, such as lakes, rivers, or hills. Other communities may be created by artificial boundaries, such as highways or buildings. Usually, communities exist because a group of people share certain qualities that make them unique. These qualities may be economic or social concerns.

In most strong communities, members also share similar interests and goals, or a **common identity**. Some communities may want to preserve the history of their neighborhoods and the appearance of their houses. Others may want to improve their neighborhoods with new housing or business developments. Many communities hold festivals or other celebrations to promote community pride.

Besides a common identity, some community members may share common cultural, racial, or religious ties. This is called a **common association**. Sometimes the people in a community such as this do not live in the same area. People with similar political or economic needs can be called a community. For example, senior citizens may identify themselves as a part of the senior community. As a community, groups of senior citizens can ask for greater

Writing About Health

Make a list of the different communities you belong to. Explain how your membership in each community affects you and your health. Then name some contributions you make to each community.

Public Health Chapter 18 **369**

Epidemic
A disease that spreads quickly

Resource
A source of supply and support

resources, sources of supply and support, to meet their unique health and social needs. Other groups of people can organize themselves into communities to make sure their needs are met.

How Is Health a Community Concern?

The health of every community member is a community concern. People with communicable diseases, such as influenza, can infect others. Communicable diseases can be passed along to others. If a disease becomes uncontrollable, it is called an epidemic. An epidemic spreads quickly and affects many people at the same time. If an epidemic occurs, communities may need to set up special clinics to care for the sick. Doctors and scientists can also start working on a vaccine, or a cure, for the disease. Sometimes, however, communities can only treat the disease and try to prevent it from spreading to others.

People with similar economic needs can be called a community.

The number of health resources available in a community affect the people who live there. When people become ill or injured, what kind of health care will they receive? Are there enough health care workers so that all patients receive the treatment they need? Sometimes a certain group of people does not get the health care they need. For example, in some areas, pregnant women may have to wait weeks to see a doctor. If the woman does not get proper care during her pregnancy, the baby may be affected. This would end up hurting the community into which the baby is born.

Economics
The way in which money and other resources are divided among community members

People dependent on drugs can have a harmful effect on the community. They may be less productive at work and more likely to become ill. They may become violent. A good community resource is a clinic that can treat drug addiction. Clinics can also treat people with mental disabilities.

How Is Economics a Community Concern?

Economics is the way in which money and other **resources** are divided among community members. All health care has a cost. Doctors and other caregivers need to be paid for their work. Supplies and equipment need to be paid for. What happens when individuals do not have enough money to pay for the care they need?

The number of people unable to afford health care is increasing. Many do not have enough health insurance. The company they work for may not offer insurance. They may be unemployed and cannot afford to buy insurance. This means they may not be able to afford needed health care for themselves or their family. By not getting proper treatment, their health problems could get worse.

Poverty is a major health concern. It puts people at risk for many serious health problems. When people cannot afford a nutritionally balanced diet, malnutrition and other health problems can result. Poverty is the main reason that some children do not get proper vaccinations and other necessary care.

What Are Some National Health Goals?

Most communities do not exist independently from the rest of the world. Federal and state governments can affect and influence many aspects of a community. They may provide communities with assistance to address problems. They may also set rules or goals that a community must meet.

Health Tip

If you need vaccinations, find out if they will be offered in a neighborhood health center or a school in your community.

Every ten years, the U.S. government sets public health goals on which to work. These public health goals include the following.

- To provide equal health care for all Americans, including children, adolescents, older people, and people with disabilities
- To increase health and quality of life for older people
- To increase the ability of all Americans to obtain the use of preventive health services

How Do Changes Affect Community Health?

All communities go through changes. Community members grow older and die, and new ones are born. Some people move in, and others move away. It is expected that the population of the United States will increase over the coming years. Most likely, the population of your community will increase, too. Communities need to change as their members change.

The number of people 65 years old or older will increase. This is because more people are living longer than they used to. This is positive for communities because it shows that people are healthier. However, it also poses special challenges. Older people have special health needs that should be considered. They can also contribute to society in different ways.

LESSON 1 REVIEW Write the answers to these questions on a separate sheet of paper. Use complete sentences.

1) Do all members of a community have to live in the same area?
2) What effect can drugs have on a community?
3) How can poverty cause health problems?
4) Which public health goal might be a concern for your parents or grandparents?
5) Why will the number of people over age 65 increase in the United States?

Lesson 2

Community Health Resources

Private resource
A group of people not associated with the government

Public resource
A health service that is run by a local, state, or national government

Volunteer resource
A group of people who donate their time to provide services

Many factors contribute to community health. Many programs and services—from local to national—can keep community members healthy.

What Are Some Local Health Resources?

Public, private, and volunteer services are available in most communities. **Public resources** are health services that are run by a local, state, or national government. Emergency Medical Service (EMS) is a public resource that provides emergency care for people with injury or illness. An ambulance with paramedics transports patients to an appropriate care facility, such as a hospital.

Other public resources include a poison control hotline, clinics, and agencies that focus on special care. Most public resources offer free or low-cost care. In addition, public programs can help pay for other types of health care.

Private resources are groups not associated with the government. Private hospitals, insurance companies, and privately run clinics are some examples of private resources. Generally, individuals pay for services received from private resources.

Volunteer resources are made up of people who donate their time to provide health information or services. There may be volunteer programs at a hospital or clinic. There are also special volunteer groups, such as the Red Cross and the American Cancer Society.

People with similar health problems or concerns may also form support groups. This may include emotional support for alcoholics and their families, single parents, and people with disabilities. For example, the Center for Independent Living (CIL) began as a local self-help group. CIL helps people who are blind or who have physical disabilities. Now the CIL is a national organization. Volunteer resources can provide health care services that might not otherwise be available.

Screening service
An exam that provides prevention, such as eye and ear exams

Statistics
Important information

Schools are another local health resource. Schools provide health education. They also provide prevention or **screening services**, such as eye and ear exams. Schools may form a partnership with a community agency to provide additional preventive services. This helps young people stay healthy and find out about problems before they occur.

What Are Some State Health Resources?

States, and sometimes cities and counties, have public health departments. These departments enforce health laws. They also provide information to help caregivers understand and treat public health problems. Health departments keep **statistics**, or important information, about public health. They might keep track of the types of diseases that have been treated in the community.

Health departments also provide free services such as vaccination and education programs. They support community clinics and substance abuse programs. They may also link people to programs that will help pay for medical expenses.

What Are Some National Health Resources?

The United States has many agencies that deal with and affect people's health. The Department of Health and Human Services (DHHS) is concerned with the health and well-being of the entire country. Many agencies operate under this department. The agencies under the Public Health Service focus on health needs and problems.

Food and Drug Administration (FDA)

The FDA tests the foods, drugs, and cosmetics we use. It requires that all food manufacturers label the ingredients on their food containers. This helps people with allergies or special diets to make smart choices about the food they eat.

Centers for Disease Control (CDC)

The CDC works to prevent the spread of disease in the United States. It keeps track of illnesses and tries to prevent an epidemic from occurring. The CDC also helps scientists develop vaccines and treatments for disease.

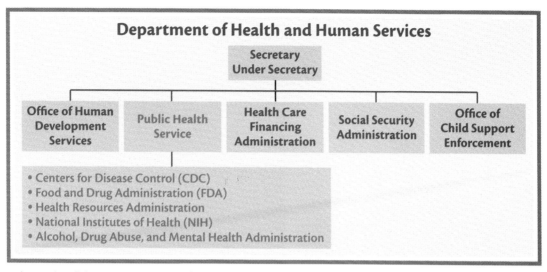

A branch of the Department of Health and Human Services, the Public Health Service is involved with health care concerns

Grant
An amount of money given

Recall
An order to return unsafe items

National Institutes of Health (NIH)

The NIH conducts research into health problems. It provides **grants**, or money, to colleges and research centers. The NIH makes sure that the information from this research is available to everyone. This research can help in the treatment of injury or illness. The research also provides information to help prevent similar problems from occurring.

A separate agency, called the Consumer Product Safety Commission (CPSC), deals with the safety of products. The CPSC can determine if a product is unsafe or damaged. It may require a **recall**, or a return of all items, to take an unsafe product off the market. The CPSC can also help people make claims against a company if they have been hurt by its product.

What Are Some International Health Resources?

Many people study how disease moves over the world. The World Health Organization (WHO) is an international organization that fights diseases that spread globally. It monitors conditions that could become world health problems. The WHO shares its information and makes decisions about distributing health resources. It can help to avoid an epidemic and save lives. The WHO promotes training and helps countries to set up effective health programs.

Most food shelf programs function through the help of volunteers of all ages.

Charitable organization
A group of people who give funds to health care programs

How Are Health Care Programs Funded?

Community health care programs get money from several sources. Federal and state governments can provide funds for prevention and special treatment programs. Other support comes from local taxes and private charities. **Charitable organizations**, such as the United Way, give funds to many different programs. These groups sometimes raise money through special fund drives. Volunteers may ask for donations from people on the street or door-to-door. They may also ask businesses to contribute. Many programs are kept running by the caring and generosity of fellow community members. These people know that the programs keep the whole community strong and healthy.

LESSON 2 REVIEW Write the answers to these questions on a separate sheet of paper. Use complete sentences.

1) What are three kinds of local health resources?

2) How do volunteer groups help communities?

3) What do public health departments keep track of?

4) How do charities help fund health care programs?

5) What does the CDC do?

Lesson 3

Community Health Advocacy Skills

Community members need to develop skills to find and manage resources. They also need to contribute to the community and be ready to change to meet new community needs.

How Do You Find Community Services?

Every community offers a range of services and programs. To participate in your community, you need to find the services that are right for you. You may have a wide range of needs, including recreational, educational, health, and housing. You will also want to know what to do in an emergency. By contacting a community service or local government, you can find out which agencies can help you.

Everyone in your community pays for the services that are offered. People pay taxes on things they buy, property they own, income they make, and business they conduct. This is how the community gets the funds that keep services going.

FINDING HEALTH RESOURCES IN YOUR COMMUNITY

Make a list of local health agencies and organizations in your community. Check with parents, librarians, teachers, and the phone book for ideas. Think about issues such as disease prevention, health promotion, and treatment programs.

Divide into pairs or teams to gather information about the agencies. Phone each agency and ask about the items below. Prepare for the interview in advance. Record the answers.

- Name, address, and phone number of the agency
- Days and hours of operation
- Types of services and health professionals on staff
- Who qualifies for services
- Fees, if any
- Whether the services are confidential
- Languages other than English spoken

Individuals play a significant role in helping to address public health problems.

Budget
A written plan for spending money

How Can People Manage Their Resources?

Sometimes it is difficult to know where to spend money first. You need food, clothing, and shelter. You need proper health care to stay healthy. You also want to do things that are fun and help you be happy. It helps to have a **budget,** or written plan for spending money. This will help you decide how much money you will need for each of your needs.

HOME HEALTH AIDE

Living at home is important for most people, especially when they are ill. Home health aides help sick people to stay at home by performing routine tasks for them. They change bed linens, prepare meals, clean, do laundry, and run errands. Aides help bathe and clean people. They may read aloud or play games with them. Under a doctor's or nurse's direction, aides give medication. Usually, health care agencies employ home health aides. Some hospitals run home care programs that employ home health aides. Home health aides should know CPR. Many states don't require training for home health aides. However, some states are developing training standards.

Expense
Something you need to spend money on

Income
The amount of money received or earned from a job

Your **income** is the amount of money that you receive or earn from a job. Your **expenses** are the things that you need to spend money on. If your expenses are greater than your income, you will go into debt. Some expenses are fixed and more important than others are. You should pay for things like food, clothing, rent or mortgage, and utility bills first. Then you can allow for some other expenses, like going to the movies or taking a vacation. It is a good idea to set aside some extra money for emergencies.

By saving money, families can have additional money available if needed for medical emergencies. They also can relax knowing that all their expenses are paid and they have money left over. They may want to use some of that money for preventive care to ensure that they stay healthy.

POPULATION CHANGES

In 1900, there were many young children 5 to 13 years old. They made up over 20 percent of the total population. At the same time, the number of people over 65 was very small. Since then, there have been many changes in the population of people 65 years and older. This is because many more people are living longer than they did years ago. Examine the graph below. How do you think this will affect communities? What health resources might people age 65 or older need?

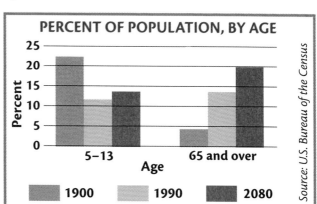

Advocacy group
An organization that works to benefit a specific group

How Can a Community Manage Resources?

A community can benefit from local, national, and international health promotion. Preventive care costs less than treatment of illness. It reduces the demands on the health care system. This allows the community to provide more care for less money.

Health promotion and disease prevention should be goals for every community. Health education is another wise use of resources. Providing information about health problems can decrease the risk of these problems occurring. Communities can make information available about diseases, violence, physical fitness, nutrition, and substance abuse.

Prevention organizations that work with specific groups are called **advocacy groups**. Advocacy groups may provide health services or help people find affordable health services. They may work with government to try to change laws or policies. Advocacy groups often focus on children, senior citizens, people with disabilities, or different ethnic communities.

How Can You Contribute to the Community?

You can do many things to get involved in and improve your community. Staying healthy is an important way to be involved in your community. When you are healthy, you are able to contribute to the community. You might choose to volunteer with a group in your community. You could also be an advocate by writing letters to lawmakers asking for change.

Health Tip

Check into volunteer opportunities that interest you and that benefit public health.

LESSON 3 REVIEW Write the answers to these questions on a separate sheet of paper. Use complete sentences.

1) How does the community pay for its services?
2) Why would a person use a budget?
3) How does preventive care save money for a community?
4) What does an advocacy group do?
5) How can people prepare for medical emergencies?

Chapter Summary

- Physical boundaries, common identity, common interests, and other ties are factors in forming a community.

- A community's health depends on the health of all its members.

- A strong community has health care programs.

- Poverty contributes to disease and other public health problems.

- Public, private, and volunteer resources provide care to members of a community.

- School health programs enhance community resources.

- State health departments help with disease control and help provide care for people who cannot afford it.

- National health resources are concerned with the health and safety of the entire country.

- Federal agencies regulate food and drugs and prevent spread of disease. Some agencies conduct medical research and determine product safety.

- Funding for community health programs comes from many sources, including national, state, and local governments.

- Charities raise and contribute money to many community causes.

- Health services are available through community programs to those who otherwise could not afford them.

- Community members help finance services by paying taxes.

- A budget can help you manage money and plan for health emergencies.

- Communities can manage resources through prevention, education, and planning.

- You can fulfill your responsibility to your community by protecting your own health and volunteering to help others.

Chapter 18 Review

Comprehension: Identifying Facts

On a separate sheet of paper, write the correct word or words from the Word Bank to complete each sentence.

WORD BANK	
advocacy group	health resources
area	identity
budget	increase
donate	poverty
Emergency Medical Service (EMS)	screening services
	taxes
epidemic	World Health Organization (WHO)
goals	

1) In a strong community, members share a common _____.

2) If a disease becomes uncontrollable, it is called an _____.

3) A community with a common association may not live in the same _____.

4) The _____ available in a community affect the people who live there.

5) _____ is the main reason that some children do not get proper vaccinations.

6) The United States has set several public health _____.

7) The U.S. population will likely _____ in the coming years.

8) _____ provides emergency care for people with injury or illness.

9) Volunteers _____ their time.

10) Schools provide _____, such as eye exams.

11) The _____ fights diseases that spread globally.

12) Community services are partly funded by _____.

13) It helps to have a _____, or written plan for spending money.

14) An _____ might try to change a law.

Comprehension: Understanding Main Ideas

Write the answers to these questions on a separate sheet of paper. Use complete sentences.

15) What is an epidemic?

16) What kind of information do state health departments provide?

17) Which Public Health Service agency would be concerned about reports of a new disease?

18) What happens if your expenses are greater than your income?

Critical Thinking: Write Your Opinion

19) Why are health promotion and prevention programs important?

20) What might a charitable organization do to fight an epidemic?

Test Taking Tip When you read true-false questions, the statement must be absolutely correct. Words like *always* and *never* tell you the question is probably false.

Chapter 19

Environmental Health

Humans are not alone on this planet. We share the earth with many living things. In recent years, there have been reports and studies on how our environment is changing. People now understand that the land, air, water, and living things are all valuable resources. These resources must be protected so that they can be available for future generations.

In this chapter, you will learn about the environment and how it affects your health. You will learn about environmental problems and what can be done to solve these problems. You will also learn how you can help to protect and preserve the environment.

Goals for Learning

▶ To explain how the environment affects health
▶ To identify causes and effects of air pollution
▶ To identify causes and effects of water and land pollution
▶ To identify actions that protect and promote a healthy environment

Lesson 1

Health and the Environment

Environment
The system that connects air, land, water, plants, and animals

All living things depend on air, water, and land for survival. The system that connects air, land, water, plants, and animals is called the **environment**. Changes in the environment affect the health of all living things.

How Does the Environment Achieve Balance?

Everything on the earth contributes to the environment. Water and minerals in the land help plants to grow. Plants give off oxygen, which animals need to breathe. Animals turn the oxygen into carbon dioxide, which plants need to survive. All parts of the environment contribute to one another. This creates a healthy balance in nature. When a part of the environment is hurt or destroyed, the balance is changed.

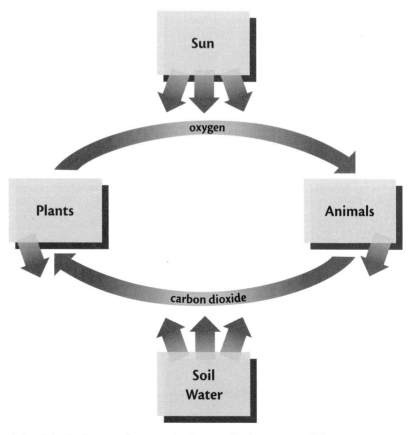

A healthy balance of nature is changed when part of the environment is hurt or destroyed.

Environmental Health Chapter 19

Healthy Subjects
Social Studies

RESULTS OF A NATURAL DISASTER

In 1998, a storm named Hurricane Mitch swept through Central America. Winds reached 180 miles per hour, and four feet of rain fell in a few days. This caused severe floods and mudslides. At least 11,000 people were killed by the storm. Millions of people were left homeless. It was the worst storm to hit the region in over one hundred years.

Many countries around the world provided aid to the people of Central America. They needed help to rebuild their homes. Also, food and drinking water were in short supply after the storm. How have people responded to natural disasters in your community?

What Disturbs the Environment's Balance?

Since many things contribute to the environment, any event can affect its balance. Natural events, such as earthquakes, volcanoes, and floods, can cause damage. But nature usually recovers from these kinds of disasters. For example, new plants grow over land damaged by a fire. In fact, the changes that natural events produce can have positive effects by renewing the environment.

Writing About Health

If a tree is cut down, what do you think will happen to the living things that used the tree as a source of food? Write your thoughts.

Human activity can also disturb the environment. Humans are the only animals who have the ability to change their environment. They do this by building houses, farming, and using the earth's resources. As more people inhabit the earth, there are more demands on the earth's resources. There is less land for plants and animals to live. This puts a strain on the environment.

Some of the resources that people use from the environment cannot be replaced. Metals and minerals are taken from the ground to make things that people use. Coal and oil are mined to heat homes or fuel vehicles, such as cars. Trees are cut down to make wood for houses and paper for writing.

Pollution
The buildup of harmful wastes in the air, land, or water

As with natural disasters, the environment will change in response to these events. Nature will try to re-create a balance. However, human activity could permanently upset the balance of nature. Land that has been mined or stripped of trees may be so damaged that it can no longer grow trees or crops. It may even be unsafe for humans to live on. People must be careful that they do not damage the environment so much that it cannot recover.

How Does the Environment Affect Our Health?

Problems in the environment cause problems for humans, too. Some effects can be noticed right away, such as a flood in an area where there are homes. Other effects may not be noticed until many years have passed. The result is **pollution**, the buildup of harmful wastes in the air, land, or water.

Over time, nature breaks down most normal wastes. However, some pollution consists of dangerous chemicals or other harmful wastes that nature cannot clean. These wastes come from human activity such as manufacturing. The products from these factories make our lives easier. However, chemicals needed to make the products can make factory workers sick. Wastes from factories and vehicles harm the air and water that humans need to live.

Pollution is a buildup of harmful wastes in the air, land, or water.

Ecology
The study of how living things are connected

How Can We Keep a Healthy Environment?

It is important to understand the problems in the environment. Then we can take steps to fix them. People now know that the earth's resources are limited. The environment is delicate and all of its parts are connected. It is important for us to preserve the environment for ourselves and future generations.

By studying the environment, you can find ways to make changes to improve it. **Ecology** is the study of how living things are connected in the environment. By careful study, humans can find ways to repair the harm they have caused to the environment.

LESSON 1 REVIEW Write the answers to these questions on a separate sheet of paper. Use complete sentences.

1) How do plants and animals contribute to one another through the environment?
2) How can mining hurt the environment?
3) How is some pollution different from normal wastes?
4) What is the study of how living things are connected?
5) What is the difference between a natural disaster and a damage to the environment by human causes?

It is important to preserve a healthy environment for future generations.

Lesson 2

Air Pollution and Health

By-product
An unwanted result

Hydrocarbon
A toxin caused by motor vehicle exhaust

Toxin
A dangerous chemical

Living things depend on clean air to breathe and live. Indoor and outdoor air pollution can affect your health. Problems in the air can also affect other living things in the environment.

Why Do We Need Clean Air?

We breathe in and out about 3,000 gallons of air every day. We use the oxygen in the air to help fuel our bodies. If the air is polluted, then **toxins**, or dangerous chemicals, will enter our bodies. Over time, these toxins can affect our lungs or other parts of our bodies. It may take years for the damage to be noticeable. If the level of toxins is high enough, there can be damaging effects in weeks or months.

What Causes Air Pollution?

The main cause of air pollution is burning. Whenever a substance is burned, harmful gases and particles are produced. The burning of fuels, such as oil, coal, or gasoline, creates smoke and exhaust. This is a **by-product**, or unwanted result, of the burning. Some other sources of air pollution are tobacco smoke, furnaces that burn waste material, and leaf burning.

Pollution can also be caused when chemicals or gases are released into the air. Usually these chemicals do not occur naturally. For example, pesticides and asbestos are made by humans.

How Does Outdoor Pollution Affect Health?

Sometimes you can see air pollution, such as smoke. Other times you cannot see pollution because the particles are so small. However, the pollutants can still harm us.

The toxins caused by the burning of motor vehicle exhaust are called **hydrocarbons**. Hydrocarbons can also enter the air when natural gas escapes into the air or when liquid fuel evaporates. Hydrocarbons can irritate your nose and throat and make your eyes water. They can also cause cancer.

Health Tip

To reduce pollution, limit how often you drive a car.

Car exhaust is one of the causes of smog.

Carbon monoxide
A pollutant caused by car exhaust

Fossil fuel
A burnable substance formed in the earth from plant or animal material

Smog
Air pollution formed by car exhaust and other pollutants

Carbon monoxide is a pollutant that car exhaust and tobacco smoke cause. It is a colorless, odorless gas that is poisonous at high levels. If you are in an area with high carbon monoxide levels, your health may be affected. You may experience headaches, dizziness, blurred vision, and fatigue. High carbon monoxide levels in areas can cause death. Over many years, carbon monoxide contributes to heart and lung disease.

When factories and power plants burn **fossil fuels**, such as coal or oil, they produce sulfur dioxide. Fossil fuels are burnable substances that are formed in the earth from plant or animal material. Sulfur dioxide irritates the eyes, throat, and lungs and makes breathing difficult.

Many big cities have a problem with **smog**. Smog is caused when there is a great deal of exhaust and other pollutants in an area. Smog makes the air look dirty and hazy. It also creates ozone, which interferes with our ability to absorb oxygen from the air. Some people may have trouble breathing in an area with a lot of smog. Ozone can cause conditions such as asthma, bronchitis, and emphysema.

Acid rain
Rain, snow, sleet, or hail with large amounts of sulfuric acid or nitric acid

Asbestos
A material made of fibers that is used as an insulator; it can cause lung cancer

Particulate
A small pollutant, such as dust, ash, or dirt

Radon
A colorless, odorless, poisonous gas that is formed underground

Particulates are small pollutants, such as dust, ash, or dirt, that hang in the air. Particulates, such as pesticides, can harm the body when they are inhaled. They can irritate the eyes, nose, throat, and lungs. Lead is a particulate that can be absorbed through the lungs or the skin. Lead enters the bloodstream and can cause lead poisoning. Symptoms include high blood pressure, loss of appetite, and brain damage.

How Does Indoor Pollution Affect Health?

If a building is insulated, or sealed up tightly, the indoor air may not circulate enough. This will cause a buildup of any toxins that may be present in the building. Glue, paint, pesticides, and household cleaners can release harmful gases into the air. Tobacco smoke can cause breathing problems for both smokers and nonsmokers.

Indoor air pollution can cause shortness of breath, coughing, and nose and throat irritation. Other symptoms are eye irritation, upset stomach, headaches, dizziness, and fatigue. Some indoor air pollutants can cause lung disease or cancer.

Houses that contain **asbestos** can be dangerous for people living in them. Asbestos is a material made of fibers. It was used as an insulator in older buildings. Asbestos fibers in the air can affect the lungs and cause cancer. Asbestos has now been banned. It is being removed from homes and other buildings.

Radon is a colorless, odorless gas that is formed underground. It can seep into houses through walls or floors. Radon is a serious indoor pollutant. It may cause lung cancer. There are methods for detecting and removing radon gas. The Environmental Protection Agency (EPA) can help if you have radon in your home.

How Can Air Pollution Hurt the Environment?

High levels of pollution can affect the entire planet. Chemicals in the air can cause **acid rain**. Rain, snow, sleet, or hail with large amounts of sulfuric acid or nitric acid is called acid rain. It damages rivers, lakes, and plant and animal life. It can destroy forests and make fish unsafe to eat.

Greenhouse effect
A gradual warming of the earth's atmosphere

Ozone layer
A region in the atmosphere that protects the earth from the sun's harmful rays

Many scientists think that the rising levels of carbon dioxide in the air will cause a **greenhouse effect**. The glass ceiling of a greenhouse traps the sun's heat inside. The pollution in the air that surrounds the earth might have the same effect. If air pollution makes the earth's temperature rise, there will be other changes in the environment. Rainfall and weather patterns could change. This could lead to food shortages and more deserts. Melting of the polar ice caps could cause the sea level to rise. Coastal areas would be flooded.

Another problem that air pollution may have caused is the reduction of the **ozone layer**. Although ozone is harmful for people to breathe, it is helpful in the atmosphere. High above the earth's surface, the ozone layer shields the earth from the sun's harmful rays. The ozone layer helps protect us from sunburn and skin cancer. In 1985, scientists discovered a hole in the ozone layer over Antarctica. The size of the hole was as big as the United States. In addition, the whole ozone layer appears to be thinning.

Scientists think that chlorofluorocarbons (CFCs) in the air may be responsible for the decrease in the ozone layer. CFCs are chemicals that were used as cooling agents in refrigerators and air conditioners. These chemicals may have risen up to the ozone layer. They react with the ozone and turn it into oxygen molecules, which cannot protect us from the sun's harmful rays.

LESSON 2 REVIEW Write the answers to these questions on a separate sheet of paper. Use complete sentences.

1) What is the main cause of air pollution?
2) What are some of the effects of smog?
3) What is one natural cause of indoor air pollution?
4) How does the ozone layer help us?
5) Name three ways air pollution harms the environment.

Lesson 3

Water and Land Pollution and Health

> **Groundwater**
> Water beneath the earth's surface

Water and land are important resources. If we use them up or pollute them, they will not be easy to replace. We have to use sparingly, or conserve, and protect our water and land to keep our environment healthy.

Why Is Water Important to Health?

People need freshwater to live. Three-quarters of the earth's surface is covered with water. However, most of that is saltwater, which cannot be used for drinking or watering crops. Freshwater is as important as air in sustaining life.

Water beneath the earth's surface is called **groundwater**. Groundwater is our biggest source of drinking water. Any toxins in the groundwater will be transferred to us when we drink it. As the world becomes more populated, freshwater is harder to find. If groundwater is not kept clean and free of pollution, we could run out of freshwater sources.

What Causes Water Pollution?

For many years, humans have dumped wastes into oceans, lakes, and rivers. Many of these wastes cannot be broken down. They remain for a long time and cause damage.

Industrial Wastes

Factories and other operations dump industrial wastes and by-products into the water. These may include substances like mercury, lead, acids, and other chemicals that cannot be broken down. These toxic substances pollute the water supply and hurt marine life.

Sewage

The wastewater from drains and toilets is usually carried off as a part of a sewage disposal system. Sewage includes human waste and other organic material. It contains bacteria and can cause disease. If raw sewage gets into groundwater, it can kill wildlife and plants. If it enters the water supply, it can harm humans. Most cities have a sewage treatment plan. Sewage is treated before it is released back into the environment.

Household Chemicals

Some products that people use every day can have harmful effects if they get into the water supply. Harsh detergents, bleach, and other cleaning products can pollute the environment. You can read the ingredient list on products to see if they contain anything harmful. The labels should also tell you how to dispose of the product safely.

Agricultural Runoff

Farmers use fertilizers and other chemicals to help grow crops. When these chemicals sink into the groundwater or streams, they can have harmful effects. As the level of chemicals builds up, they can be harmful to fish, animals, and people.

Oil Leaks and Spills

All over the world, people use oil and the products that come from it, such as gasoline. Oil is often shipped across the oceans. Sometimes the oil leaks or spills. The spilled oil does not mix with the water. Instead, it sits on top of the water like a thick skin. The oil can kill wildlife and leave behind dangerous toxins. Oil spills are difficult to clean up. The effects of an oil spill can last for years.

Oil spills kill wildlife and leave behind toxins that can affect humans.

Famine
A shortage of food

How Can Freshwater Be Conserved?

Reducing the amount of pollutants in our water is important. By protecting the water we have, we can avoid a shortage later. We can look for cleaner ways to dispose of waste, chemicals, and other products. Another way to protect freshwater is to conserve it. Here are some ways you can conserve water:

- Turn off the water in the shower while you lather and clean. Limit your time in the shower to five minutes.
- Fill the tub only halfway when you take a bath.
- Turn off the water when you are not using it. Don't let it run while you brush your teeth.
- Wait to run the clothes washer or dishwasher until there is a full load.

How Can Land Pollution Affect Your Health?

The number of people on the earth keeps growing, and the amount of open space keeps shrinking. We need to manage carefully the land we have so that we can meet our needs. We also must find solutions to problems like overpopulation and ways to manage solid waste.

Population Growth

There are nearly six billion people on the earth. That number may grow to nine billion over the next fifty years. A larger population needs more food. There is, however, only a limited amount of land. How can we grow more crops on less land? Many countries already have problems with **famine**, or a shortage of food. This can happen when there is not enough rain to grow crops.

A larger population would also need more cars and other products. This can cause problems because the cars and factories that make products also cause pollution. Increased pollution can cause more problems for the environment.

Health Tip

Reduce your waste by not buying things that you don't need.

Landfill
A place where waste is buried between layers of earth

Solid Waste

Getting rid of the waste we create is a big job. Solid waste is usually dumped in a **landfill**. A landfill is a place where waste is buried between layers of earth. When landfills get too full, some of the trash may be burned. However, this pollutes the air. Sometimes toxic waste is dumped in a landfill. If it is not disposed of properly, the toxic waste can seep into the ground and poison the water.

How Can We Reduce the Stress on the Environment?

Every person must work at preserving the environment and conserving our resources. Here are some actions you can take to keep the earth a healthy place to live:

- Buy products that can be used again and again.
- Look for products with the recycle symbol.
- Buy products that have the least packaging.

LESSON 3 REVIEW Write the answers to these questions on a separate sheet of paper. Use complete sentences.

1) What is the source for most of our drinking water?
2) Why would the effects of an oil spill last for years?
3) How can you conserve water in the shower?
4) What might happen when a landfill gets too full?
5) How does a larger population strain the earth's resources?

Recycling paper and aluminum cans reduces the stress on the environment.

Lesson 4

Promoting a Healthy Environment

Protecting the environment is everyone's job. Governments, industries, and individuals can work together to make the earth a healthy place to live.

How Can Governments Protect the Environment?

The U.S. government has passed laws to protect the environment and people's health. The Clean Air Act was passed in 1970, and changes have been made to improve it since then. It sets limits on the level of pollutants. There are also laws about water safety and the disposal of solid and toxic wastes.

The government uses these laws as guidelines for keeping the environment clean. For example, it might fine a factory for polluting the air or water.

Local governments and communities can also work on protecting the environment. They can set up programs and educate people on how they can contribute. Sometimes cities will issue a smog alert. Everyone should try to avoid or reduce driving on those days. People who have health problems should stay indoors because the air may be unhealthy.

Careers

ENVIRONMENTAL PROTECTION WORKER

If you are concerned about the environment, you could become an environmental protection worker. Environmental protection workers support good health and environmental practices. Using computers and other equipment, they check for poisons in samples of air, water, and soil. They also make sure laws about health standards are followed. They may check sewage treatment plants to make sure the water is cleaned properly. Environmental protection workers also check the air for dangerous gases. Training and certification vary by state. Specialties such as those working with asbestos require training and certification from the Occupational Safety and Health Administration (OSHA).

Hoover Dam supplies water and electric power for a large area of the Pacific Southwest.

What Are Some Future Environmental Solutions?

Scientists and other researchers are trying to find new ways of protecting the environment. An alternative source of energy that does not pollute the environment would keep the air and the water clean. Solar energy and wind energy are some examples.

The government and the automobile industry are also working on more efficient cars that put out less pollution. For example, California is working to increase the number of electric cars on its roads. Electric cars put out little or no pollution. They may be more energy efficient than gasoline-powered cars. Alternative fuels are another possibility for automobile improvements. Natural gas and hydrogen are some possibilities for the future.

Engineers are also working on ways to grow more crops in a smaller area. Perhaps some of the farmland that is saved can be changed back to wetland or forest.

DISAPPEARING WETLANDS

Wetlands create a rich home for plants and wildlife in our environment. They prevent floods because they can hold a great deal of water. Wetlands also absorb pollutants and help keep the water clean. One hundred years ago, wetlands covered much of the United States, especially in the Midwest. Today, about half of all U.S. wetlands have been lost. Farms, houses, or other buildings have replaced them. Wetlands are also sensitive to changes in the environment. Some groups of people are trying to raise awareness of the importance of wetlands. Some wetland areas are now protected. Researchers are also trying to find ways to create new wetlands to replace the ones that were lost.

How Can Individuals Protect the Environment?

Individuals can make a difference. Everyone can take actions that will improve the environment. Here are some things you can do:

- Walk, ride a bike, or take mass transit instead of driving.
- Turn off the lights when you leave a room.
- Set air conditioners no lower than 75 degrees in summer.
- Keep the furnace set no higher than 68 degrees in winter.
- Recycle newspapers, cans, bottles, plastic, and paper. Reuse items like paper clips.
- Select products that contain the least amount of packaging. Larger containers usually hold more product with less packaging.
- Purchase consumable items, such as detergent, in refillable containers.
- Find alternatives to toxic products around your home. For example, use baking soda and water instead of oven cleaners and harsh cleansers. Wash windows with vinegar and water.

Action for Health

RECYCLING

Recycling means reusing materials instead of buying new ones. You can make a difference in your community's environment by recycling. It saves natural resources and also reduces solid waste. Newspapers, cardboard, aluminum cans, glass, and some paper and plastics can be recycled. Find out if your community has a recycling program, what materials it takes, and how it works. Organize a drive to collect materials and take them to a recycling center.

LESSON 4 REVIEW Write the answers to these questions on a separate sheet of paper. Use complete sentences.

1) What would you try to avoid during a smog alert?
2) Name some alternative energy sources.
3) Why might an electric car be better for the environment?
4) How does the government help protect the environment?
5) Name five ways individuals can help protect the environment.

Chapter Summary

- All parts of the environment contribute to one another.

- Humans are the only animals capable of changing their environment.

- The environment recovers from natural events and disasters. These events sometimes renew the environment.

- Human activity could damage the environment and upset its balance.

- Pollution can endanger human health and life as well as other forms of life.

- Understanding environmental problems is the first step in finding solutions.

- Burning is the main cause of outdoor air pollution.

- Pollutants, such as hydrocarbons, smog, ozone, and carbon monoxide, can cause harm to humans.

- Indoor air pollutants can be a problem in a building without good circulation.

- Indoor air pollutants include gases and fumes from household items, tobacco smoke, radon, and asbestos.

- The greenhouse effect, thinning of the ozone layer, and acid rain are some global air pollution problems.

- Freshwater is an important and limited resource that is necessary for human life. Most freshwater comes from groundwater.

- Industrial wastes, sewage, oil leaks, agricultural runoff, and toxic chemicals can cause water pollution.

- The growing world population may cause many environmental problems, including increased land pollution.

- The U.S. government has passed laws that help maintain a healthy environment.

- Alternative energy and fuel sources will help to protect our environment.

- Individuals can take actions to protect the environment.

Chapter 19 Review

Comprehension: Identifying Facts

On a separate sheet of paper, write the correct word or words from the Word Bank to complete each sentence.

WORD BANK	
bacteria	landfill
balance	ozone
carbon dioxide	pollutant
Clean Air Act	pollution
ecology	skin cancer
famine	solar energy
groundwater	toxin

1) Plants give off oxygen, which animals turn back into _____.

2) Natural events and human activity can upset the _____ of nature.

3) Harmful wastes that cannot be broken down by nature create _____.

4) _____ is the study of how living things are connected.

5) A _____ is a dangerous chemical.

6) _____ is created by smog.

7) Radon gas is an indoor _____.

8) The ozone layer helps protect us against sunburn and _____.

9) Most of our drinking water comes from _____.

10) Sewage treatment kills _____ that can cause disease.

11) A country may experience a _____ if it does not have enough crops to feed all the people.

12) The _____ sets limits on the level of pollutants allowed in the United States.

13) _____ is an alternative energy source that does not pollute the environment.

14) A _____ is a place where waste is buried between layers of earth.

Comprehension: Understanding Main Ideas

Write the answers to these questions on a separate sheet of paper. Use complete sentences.

15) What is a by-product of burning fossil fuels?

16) Which pollutant has been used as insulation?

17) What air pollution problem could cause global temperatures to rise?

18) Where are most solid wastes stored?

Critical Thinking: Write Your Opinion

19) If acid rain killed all the fish in a lake, how might other animals in the area be affected?

20) What would you do if you wanted to get rid of some old cleaning supplies that you found in the basement?

Test Taking Tip When taking a short-answer test, first answer the questions you know for sure. Then go back to spend time on the questions about which you are less sure.

Deciding for Yourself

The Clean Water Supply

Clean water is important for our health. We need clean water for cooking, washing, and drinking. Getting a clean water supply is important for all communities. Some communities have difficulty providing enough water for everyone. For example, in Los Angeles, water is piped in from hundreds of miles away. Other communities might have to dig very deep wells to find groundwater.

As the population grows, the demand for clean water is becoming greater. Communities that ship out water to other places may find that they need it for themselves. Pollution can also make it harder to find clean water.

There has been a population boom in Arizona. Many people want to move there because the weather is usually warm and dry. People who are elderly or sick may move there to improve their health. But much of Arizona is dry and desert-like. The population needs water to live. New lawns and golf courses that have been built need to be watered to be maintained. This creates a water shortage problem. People in Arizona and in other places where there is a water shortage need to decide how to conserve water. What is the best use of water? What uses are wasteful? This is a problem that will continue to grow as the population grows and pollution problems get worse.

Questions

1) Before moving into a community, would you consider where it gets its water?

2) What would you describe as a poor use of water?

3) How can we keep our water clean?

4) What are some ways to conserve water?

Unit Summary

- The health care industry includes many people in different jobs and many products and services.

- Prevention and taking care of colds and minor illnesses are wise forms of self-care.

- Individuals can pay for health care costs out-of-pocket or through private insurance or managed care.

- The U.S. government protects consumers with some basic rights. Many health care organizations also set forth rights for patients.

- Consumers should read product labels and consult a medical professional to decide whether to buy a product.

- Public, private, and volunteer resources provide care to members of a community.

- National health resources are concerned with the health and safety of the entire country.

- Some funding for community health programs comes from national, state, and local governments. Community members help finance services by paying taxes.

- You can fulfill your responsibility to your community by protecting your own health and volunteering to help others.

- Understanding environmental problems is the first step in finding solutions.

- Pollutants, such as hydrocarbons, smog, ozone, and carbon monoxide, can cause harm to humans.

- The greenhouse effect, thinning of the ozone layer, and acid rain are some global air pollution problems.

- The growing world population may cause many environmental problems, including increased land pollution.

- Individuals can take actions to protect the environment.

UNIT 7 REVIEW

Comprehension: Identifying Facts

On a separate sheet of paper, write the correct word or words from the Word Bank to complete each sentence.

WORD BANK	
Better Business Bureau	poverty
carbon dioxide	prevention
Clean Air Act	rights
donate	skin cancer
health professional	solar energy
managed care	specialist
pollution	taxes

1) _____ includes keeping a healthy body weight and not using tobacco or alcohol.

2) A person should seek a _____ if the flu or a fever lasts more than a week.

3) In _____, you may have a limited choice of doctors.

4) Your primary care physician may refer you to a _____ for a certain illness.

5) The American Hospital Association established a list of _____ for patients.

6) If a manufacturer or provider does not respond to a consumer complaint, you could contact the _____.

7) _____ is the main reason that some children do not get proper vaccinations.

8) Volunteers _____ their time.

9) Community services are partly funded by _____.

10) Plants give off oxygen, which animals turn back into _____.

11) Harmful wastes that cannot be broken down by nature create _____.

12) The ozone layer helps protect us against sunburn and _____.

13) The _____ sets limits on the level of pollutants allowed in the United States.

14) _____ is an alternate energy source that does not pollute the environment.

Comprehension: Understanding Main Ideas

Write the answers to these questions on a separate sheet of paper. Use complete sentences.

15) If you have a low income, how could you pay for health care?
16) What is an epidemic?
17) What is a by-product of burning fossil fuels?
18) What air pollution problem could cause global temperatures to rise?

Critical Thinking: Write Your Opinion

19) Why do we need a Consumer Bill of Rights?
20) Why are health promotion and prevention programs important?

Appendix A: Nutrition Tables

Table A.1. The Six Essential Nutrient Classes

NUTRIENT	BEST FOOD SOURCES	WHY THEY ARE NEEDED
Protein	Cheese, eggs, fish, meat, milk, poultry, soybeans, nuts, dry beans, and lentils	To promote body growth; to repair and maintain tissue
Carbohydrate	Bread, cereal, flour, potatoes, rice, sugar, dry beans, fruit	To supply energy; to furnish heat; to save proteins to build and regulate cells
Fat	Butter, margarine, cream, oils, meat, whole milk, nuts, avocado	To supply energy; to furnish heat; to save proteins to build and regulate cells; to supply necessary fat-soluble vitamins and other nutrients
MINERALS		
Calcium	Milk, cheese, leafy green vegetables, oysters, almonds	To give rigidity and hardness to bones and teeth; for clotting of blood, osmosis, action of heart and other muscles, and nerve response
Iron	Meats (especially liver), oysters, leafy green vegetables, legumes, dried apricots or peaches, prunes, raisins	To carry oxygen in the blood
Iodine	Seafood, iodized salt	To help the thyroid gland regulate cell activities for physical and mental health
VITAMINS		
Vitamin A	Whole milk, cream, butter, liver, egg yolk, leafy green vegetables, dark yellow fruits and vegetables	To promote health of epithelial tissues; for health of eyes and development of teeth
Thiamin	Present in many foods, abundant in few; pork, some animal organs, some nuts, whole grains, yeast, dry beans, and peas	To promote healthy nerves, appetite, digestion, and growth; for metabolism of carbohydrates
Riboflavin	Milk, glandular organs, lean meats, cheese, eggs, leafy green vegetables, whole grains	To make for better development, greater vitality, freedom from disease, metabolism of carbohydrates, fats, and proteins
Niacin	Lean meats, liver, poultry, peanuts, legumes, yeasts	To promote good digestion, healthy skin, and a well-functioning nervous system
Vitamin C	Citrus fruits, strawberries, tomatoes, broccoli, cabbage, green peppers	To enhance iron absorption; for deposit of intercellular cement in tissues and bone
Vitamin D	Milk, salmon, tuna, action of sun	To help absorb and use calcium and phosphorus
WATER		
	Drinking water, foods	To supply body fluids, regulate body temperature

Table A.2. Fiber Content of Selected Foods

GRAINS

	Serving size	Dietary Fiber (g)
Bread, white	1 slice	0.6
Bread, whole wheat	1 slice	1.5
Oat bran, dry	1/3 cup	4.0
Oatmeal, dry	1/3 cup	2.7
Rice, brown, cooked	1/2 cup	2.4
Rice, white, cooked	1/2 cup	0.8

FRUITS

Apple, with skin	1 small	2.8
Apricots, with skin	4 fruit	3.5
Banana	1 small	2.2
Blueberries	3/4 cup	1.4
Figs, dried	3 fruit	4.6
Grapefruit	1/2 fruit	1.6
Pear, with skin	1 large	5.8
Prunes, dried	3 medium	1.7

VEGETABLES

Asparagus, cooked	1/2 cup	1.8
Broccoli, cooked	1/2 cup	2.4
Carrots, cooked, sliced	1/2 cup	2.0
Peas, green, frozen, cooked	1/2 cup	4.3
Potato, with skin, raw	1/2 cup	1.5
Tomato, raw	1 medium	1.0

LEGUMES

Kidney beans, cooked	1/2 cup	6.9
Lima beans, canned	1/2 cup	4.3
Pinto beans, cooked	1/2 cup	5.9
Beans, white, cooked	1/2 cup	5.0
Lentils, cooked	1/2 cup	4.7
Peas, blackeye, canned	1/2 cup	4.7

Table A.3. Calcium and Fat Content of Dairy Products*

PRODUCT	CALORIES	FAT (grams)	CALCIUM (milligrams)
Skim milk (also called nonfat or fat free)			
Plain	86	0	301
Chocolate	144	1	292
1% milk (also called lowfat or light)			
Plain	102	2½	300
Chocolate	158	2½	288
2% milk (now called reduced fat, not lowfat)			
Plain	121	5	298
Chocolate	179	5	285
Whole milk			
Plain	150	8	290
Chocolate	209	8	280
Buttermilk (lowfat)	100	2	300
Buttermilk (whole)	150	8	300
Sweetened condensed milk	246	7	217

*All nutrition information is based on a one-cup serving except sweetened condensed milk, based on a quarter-cup serving

Appendix B: Fact Bank

FRAME SIZE

To determine your frame size:

1. Extend your arm in front of your body bending your elbow at a ninety-degree angle to your body (your arm is parallel to your body).
2. Keep your fingers straight and turn the inside of your wrist to your body.
3. Place your thumb and index finger on the two prominent bones on either side of your elbow, measure the distance between the bones with a tape measure or calipers.
4. Compare to the medium frame chart below. Select your height based on how tall you are barefoot. If you are below the listed inches, your frame is small. If you are above, your frame is large.

ELBOW MEASUREMENTS FOR MEDIUM FRAME

Height in 1" Heels	Elbow Breadth
Men	
5'2"–5'3"	2½"–2⅝"
5'4"–5'7"	2⅝"–2⅞"
5'8"–5'11"	2¾"–3"
6'0"–6'3"	2¾"–3⅛"
6'4"	2⅞"–3¼"
Women	
4'10"–4'11"	2¼"–2½"
5'0"–5'3"	2¼"–2½"
5'4"–5'7"	2⅜"–2⅝"
5'8"–5'11"	2⅜"–2⅝"
6'0"	2½"–2¾"

Table B.1. Top 10 Fat-Blasting Exercises

ACTIVITY	CALORIES BURNED (per 30 minutes)	ACTIVITY	CALORIES BURNED (per 30 minutes)
Bicycling, vigorous (15 MPH)	340	Spinning class (indoor cycling)	312
Jogging (10- to 12-minute miles)	340 to 272	Jumping rope, slowly	272
		Tennis, singles	272
Swimming, vigorous	340	Hiking, uphill	238
Cross-country ski machine	323	Inline skating	238
		Walking, uphill (3.5 MPH)	204

Table B.2. Recommended Height and Weight for Women

Height Feet Inches	Small Frame	Medium Frame	Large Frame
4'10"	102–111	109–121	118–131
4'11"	103–113	111–123	120–134
5'0"	104–115	113–126	122–137
5'1"	106–118	115–129	125–140
5'2"	108–121	118–132	128–143
5'3"	111–124	121–135	131–147
5'4"	114–127	124–138	134–151
5'5"	117–130	127–141	137–155
5'6"	120–133	130–144	140–159
5'7"	123–136	133–147	143–163
5'8"	126–139	136–150	146–167
5'9"	129–142	139–153	149–170
5'10"	132–145	142–156	152–173
5'11"	135–148	145–159	155–176
6'0"	138–151	148–162	158–179

Weights at ages 25–59 based on lowest mortality.

Weight in pounds according to frame (in indoor clothing weighing 3 lb.; shoes with 1" heels).

Table C.3. Recommended Height and Weight for Men

Height Feet Inches	Small Frame	Medium Frame	Large Frame
5'2"	128–134	131–141	138–150
5'3"	130–136	133–143	140–153
5'4"	132–138	135–145	142–156
5'5"	134–140	137–148	144–160
5'6"	136–142	139–151	146–164
5'7"	138–145	142–154	149–168
5'8"	140–148	145–157	152–172
5'9"	142–151	148–160	155–176
5'10"	144–154	151–163	158–180
5'11"	146–157	154–166	161–184
6'0"	149–160	157–170	164–188
6'1"	152–164	160–174	168–192
6'2"	155–168	162–178	172–197
6'3"	158–172	167–182	176–202
6'4"	162–176	171–187	181–207

Weights at ages 25–59 based on lowest mortality.

Weight in pounds according to frame (in indoor clothing weighing 5 lb.; shoes with 1" heels).

Leading Causes of Death

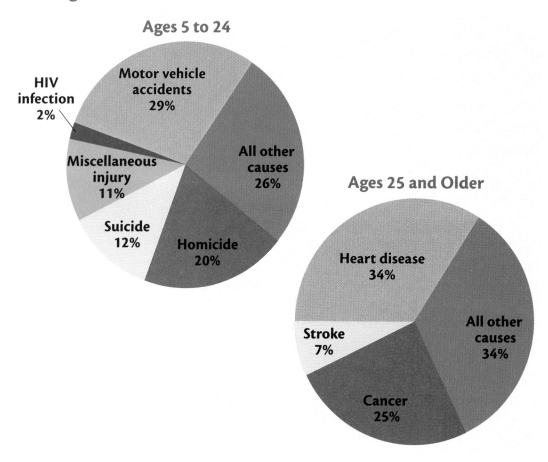

Table B.4. The Most Common Places for Cancer in Men and Women

Men	Women
Prostate Gland 184,500	Breast 178,700
Lung 91,400	Lung 80,100
Colon & Rectum 64,600	Colon & Rectum 67,000
Urinary Bladder 39,500	Uterus 36,100
Non-Hodgkin's Lymphoma 31,100	Ovary 25,400
Skin—Melanoma 24,300	Non-Hodgkin's Lymphoma 24,300
Mouth 20,600	Skin—Melanoma 17,300
Kidney 17,600	Urinary Bladder 14,900
Blood 16,100	Pancreas 14,900
Stomach 14,300	Cervix 13,700
All Sites 627,900	All Sites 600,700

Source: American Cancer Society, Inc., 1998.

The ABCDs of Melanoma

Melanoma is usually curable if you find it early. Follow this A-B-C-D self-examination guide adapted from the American Academy of Dermatology:

■ *A is for asymmetry*—Symmetrical round or oval growths are usually benign. Look for irregular shapes where one half is a different shape than the other half.

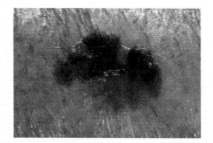

■ *B is for border*—Irregular, notched, scalloped, or vaguely defined borders need to be checked out.

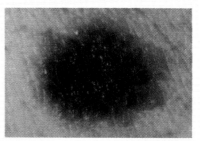

■ *C is for color*—Look for growths that have many colors or an uneven distribution of color. Generally, growths that are the same color all over are benign.

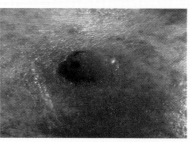

■ *D is for diameter*—Have your doctor check out any growths that are larger than 6 millimeters, about the diameter of a pencil eraser.

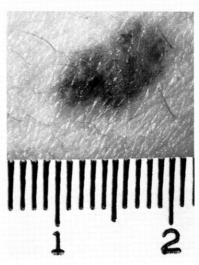

Glossary

Absorption—The moving of nutrients from the digestive system to the circulatory system (p. 140)
Acid rain—Rain, snow, sleet, or hail with large amounts of sulfuric acid or nitric acid (p. 391)
Acne—Clogged skin pores that causes pimples or blackheads (p. 55)
Acquired disease—A disease caused by infection or human behavior (p. 232)
Adapt—Change (p. 89)
Additive—A chemical added to food to make it better in some way (p. 171)
Adolescent—A child between the ages of 13 and 17 (p. 69)
Adrenal gland—The endocrine gland that releases several hormones (p. 46)
Adrenaline—The hormone that increases certain body functions (p. 46)
Advocacy group—An organization that works to benefit a specific group (p. 380)
Aerobic exercise—An exercise that raises the heart rate (p. 59)
Aggression—Any act that is meant to harm someone (p. 101)
AIDS—Acquired immunodeficiency syndrome, a disorder of the immune system (p. 243)
Alateen—A support group for teens (p. 218)
Alcohol—A chemical that depresses the central nervous system (p. 198)
Alcohol abuse—Drinking too much alcohol or drinking too frequently (p. 199)
Alcoholic beverage—A drink that contains alcohol (p. 198)
Alcoholics Anonymous (AA)—An organization that helps people live alcohol-free lives (p. 201)
Alcoholism—A disease in which a person is dependent on the use of alcohol (p. 200)
Amino acid—The smaller units of protein (p. 144)
Amphetamine—A synthetic stimulant (p. 204)
Anabolic steroid—A synthetic drug that resembles the hormone testosterone (p. 207)
Anorexia—An eating disorder in which a person chooses not to eat (p. 112)
Antibiotic—A drug used to fight bacterial infections (p. 187)
Antibody—A protein that is stimulated by the immune system to fight disease (p. 188)
Antihistamine—A medicine used for treating allergy symptoms (p. 187)

Anus—The opening through which solid wastes leave the body (p. 33)
Anxiety—An unpleasant feeling like fear but without reasons that are clear (p. 106)
Aorta—The largest artery in the body (p. 38)
Arteriosclerosis—A chronic disease in which the walls of the arteries thicken (p. 256)
Artery—A blood vessel that carries blood away from the heart (p. 38)
Arthritis—A group of diseases marked by swollen and painful joints (p. 268)
Asbestos—A material made of fibers that is used as an insulator; it can cause lung cancer (p. 391)
Asthma—A disease that affects the lungs, making it difficult to breathe (p. 263)
Atherosclerosis—A narrowing of the arteries due to a buildup of fat (p. 256)

Benign tumor—A mass of cells that are not harmful (p. 259)
Bloated—swollen (p. 162)
Blood pressure—The force of the blood against the walls of the arteries when the heart beats (p. 59)
Body image—The way each person sees himself or herself (p. 112)
Body system—A group of organs that work together to carry out a certain job (p. 25)
Bogus—Nongenuine (p. 356)
Bond—An emotional feeling of closeness (p. 119)
Budget—A written plan for spending money (p. 378)
Bulimia—An eating disorder in which a person eats large amounts of food and then vomits (p. 113)
By-product—An unwanted result (p. 389)

Caffeine—A stimulant found in coffee, tea, chocolate, and some soft drinks (p. 204)
Calorie—The unit used to measure the amount of energy in foods (p. 141)
Cancer—A group of diseases marked by the abnormal and harmful growth of cells (p. 259)
Capillary—A tiny blood vessel (p. 36)
Carbohydrate—A nutrient needed mostly for energy (p. 142)

414 Glossary   Discover Health

Carbon monoxide—A pollutant caused by car exhaust (p. 390)
Cardiac arrest—A condition in which the heart has stopped beating and there is no pulse (p. 315)
Cardiopulmonary resuscitation (CPR)—An emergency procedure for cardiac arrest (p. 315)
Cardiovascular—Relating to the heart and blood vessels (p. 188)
Cardiovascular disease—A disease of the heart and blood vessels (p. 255)
Cardiovascular fitness—The condition of the heart, lungs, and blood vessels (p. 59)
Cardiovascular medicine—A drug used for the heart and blood vessels (p. 188)
Caries—Cavities in the teeth (p. 56)
Cataract—A clouding of the lens of the eye (p. 267)
Cell—The basic unit that makes up your body (p. 24)
Cell membrane—The outer wall of a cell (p. 24)
Central nervous system—The brain and the spinal cord (p. 42)
Cerebellum—The lower part of the brain, which controls balance and coordination (p. 40)
Cerebrum—The top part of the brain, which controls thinking (p. 40)
Charitable organization—A group of people who give funds to health care programs (p. 376)
Chemotherapy—A cancer treatment that uses drugs to kill cancer cells (p. 261)
Child abuse—An action that harms a child (p. 73)
Chlamydia—A sexually transmitted disease that often has no symptoms (p. 249)
Cholesterol—A waxy, fatlike substance found in animal products (p. 145)
Chronic—Lasting (p. 247)
Chronic disease—A disease that lasts a long time (p. 268)
Circulatory system—The body system that pumps blood through the body (p. 36)
Cocaine—A dangerous and illegal stimulant drug made from the coca plant (p. 204)
Codeine—A prescription medicine that relieves severe pain (p. 191)
Commitment—Love, dedication; a pledge of trust (p. 67)
Common association—Similar cultural, racial, or religious ties (p. 369)
Common identity—Similar interests or goals (p. 369)
Communicable—Able to be passed from one person to another (p. 245)
Community—A group of people who live in the same place or have common interests (p. 369)
Competition—Many companies trying to sell similar products (p. 359)
Complete protein—A protein that has all nine essential amino acids (p. 144)
Compression—The act of pressing down (p. 317)
Compromise—An agreement in which both sides give in a little (p. 95)
Conflict—A disagreement or difference of opinion (p. 325)
Consumer—A person who buys goods and services (p. 354)
Consumer advocate—A person or group that helps consumers correct problems (p. 361)
Contact poisoning—A poison that comes in contact with the skin (p. 318)
Contaminate—Infect by contact with germs or toxins (p. 172)
Contract—Shorten (p. 28)
Convulsion—A drawing tightly together and relaxing of a muscle (p. 203)
Cope—Deal with a problem (p. 102)
Count—The number of items in a package (p. 355)
Crack cocaine—A form of cocaine that is smoked (p. 204)
Cycle—A repetition (p. 326)
Cytoplasm—The jelly-like material inside the cell membrane (p. 24)

D

Daily Values—The part of the food label that tells about the percent of nutrients in the food, based on 2,000 calories per day (p. 169)
Decibel—A unit that measures sound (p. 57)
Decongestant—A medicine that opens lung and nasal passages (p. 187)
Deductible—The initial amount a patient must pay before insurance covers health care costs (p. 352)
Defective—Not working properly (p. 354)
Depressant—A drug that slows down the central nervous system (p. 198)
Depressed—Extremely sad (p. 71)
Depression—Extreme sadness (p. 92)
Dermatologist—A doctor who takes care of skin (p. 56)
Dermis—The middle layer of the skin (p. 26)
Designated driver—The person in a group who will not drink alcohol and will drive the group home (p. 200)
Designer drug—An illegal manufactured drug that is almost the same as a legal drug (p. 208)

Discover Health Glossary **415**

Detoxification—The removal of a drug from the body (p. 217)
Diabetes—A disease in which the body does not make enough insulin (p. 188)
Digestion—The breaking down of food into nutrients (p. 140)
Digestive system—The body system that breaks food down (p. 32)
Discipline—Correct behavior (p. 68)
Disease—A disorder of normal body function (p. 232)
Disinfectant—A chemical used to prevent the spread of disease (p. 198)
Divorce—The end of a marriage (p. 71)
Drug—A substance that changes the way the mind or body works (p. 186)
Drug abuse—The improper use of legal or illegal drugs (p. 213)
Drug dependence—The need for a drug that results from the frequent use of that drug (p. 213)

Earthquake—A shaking of the rocks that make up the earth's crust (p. 297)
Eating disorder—A health problem in which a person loses control over eating patterns (p. 112)
Ecology—The study of how living things are connected (p. 388)
Economics—The way in which money and other resources are divided among community members (p. 371)
Elevate—Raise (p. 307)
E-mail—Messages sent and received over the Internet (p. 294)
Emergency kit—A collection of items that are useful in almost any kind of emergency (p. 295)
Emergency Medical Service (EMS)—An intercommunity emergency system that sends out fire, police, and ambulances by dialing 911 or 0 for operator (p. 305)
Emotional abuse—A mistreatment through words, gestures, or lack of affection (p. 73)
Emotions—Feelings (p. 86)
Emphysema—A serious disease of the lungs that causes difficulty in breathing (p. 195)
Endocrine system—The body system that uses chemicals to send and receive messages (p. 44)
Enrichment—Adding extra nutrients to a food (p. 171)
Environment—The system that connects air, land, water, plants, and animals (p. 385)

Epidemic—A disease that spreads quickly (p. 370)
Epidermis—The outer layer of skin that you can see (p. 26)
Epilepsy—A chronic disease that is caused by disordered brain activity (p. 270)
Esophagus—The tube that connects the throat and the stomach (p. 32)
Ethyl alcohol—A kind of alcohol found in beer, wine, and hard liquors (p. 200)
Excretory system—The body system that rids the body of waste and extra water (p. 32)
Expense—Something you need to spend money on (p. 379)
Extended family—A family who includes many people from different generations (p. 67)

Family life cycle—The changes in a family over time (p. 67)
Famine—A shortage of food (p. 395)
Febrile seizure—A seizure that is common in young children (p. 272)
Fiber—The parts of food that the body cannot digest (p. 143)
Firearm—A handgun or rifle (p. 287)
First aid—The immediate care given to a sick or an injured person before professional help arrives (p. 305)
Flood—A condition in which a body of water overflows and covers land that is not usually under water (p. 300)
Flood warning—A situation in which flooding has occurred (p. 300)
Flood watch—A situation in which flooding is possible (p. 300)
Food and Drug Administration (FDA)— A government agency that oversees the testing and sale of medicines (p. 189)
Food Guide Pyramid—A chart that can be used to choose a healthy diet (p. 151)
Fortification—Adding a nutrient that a food lacks (p. 171)
Fossil fuel—A burnable substance formed in the earth from plant or animal material (p. 390)
Fracture—A cracked or broken bone (p. 307)
Frostbite—A tissue injury causing the tissue to freeze due to overexposure to cold temperatures (p. 310)
Frustration—An unpleasant feeling that happens when goals are blocked (p. 101)

Gene—Parts of a cell that are passed from parent to child (p. 232)
Generic—Nonbrand-name (p. 356)
Genital herpes—A sexually transmitted chronic infection (p. 247)
Gestational diabetes—Diabetes that develops during pregnancy (p. 266)
Gland—A group of cells that produces a special substance to help the body work (p. 26)
Glaucoma—An eye disease in which pressure damages the main nerve of the eye (p. 267)
Gonorrhea—A sexually transmitted disease that often has no symptoms (p. 248)
Good Samaritan Laws—The laws that protect people who assist victims in an emergency (p. 306)
Gossip—The spreading of rumors, usually untrue, about people (p. 128)
Grand mal seizure—A seizure that affects a person's motor skills (p. 270)
Grant—An amount of money given (p. 375)
Greenhouse effect—A gradual warning of the earth's atmosphere (p. 392)
Grief—A mixture of painful emotions that result from loss (p. 91)
Groundwater—Water beneath the earth's surface (p. 393)
Guilt—An emotion felt when a person does something wrong (p. 90)

Hallucination—A distortion of the senses caused by mental disease or drugs (p. 205)
Hallucinogen—A drug that confuses the way the brain processes information (p. 202)
Hassle—A small, annoying event or problem (p. 104)
Hazard—A danger (p. 288)
Health care facility—A place where people go for medical, dental, and other care (p. 350)
Health insurance—A plan that pays all or part of medical costs (p. 352)
Heart attack—A condition in which the blood supply to the heart is greatly reduced or stopped (p. 257)
Heart rate—The number of times the heart pumps blood each minute (p. 59)
Heat exhaustion—A condition resulting from physical exertion in very hot temperatures (p. 309)
Heatstroke—A condition resulting from being in high heat too long (p. 309)
Heimlich maneuver—Firm, upward abdominal thrusts that force out foreign objects blocking an airway (p. 312)
Heroin—A dangerous and illegal narcotic drug (p. 202)
HIV—Human immunodeficiency virus, the virus that causes AIDS (p. 243)
HIV negative—Not having HIV in the blood (p. 244)
HIV positive—Having HIV in the blood (p. 244)
Hormone—A chemical messenger produced by a gland (p. 45)
Hospice—A long-term care facility for people who are dying (p. 351)
Hurricane—A tropical storm that forms over the ocean (p. 299)
Hurricane warning—A situation in which a hurricane has reached land (p. 299)
Hurricane watch—A warning to prepare for a hurricane (p. 299)
Hydrocarbon—A toxin caused by motor vehicle exhaust (p. 389)
Hypertension—High blood pressure (p. 255)
Hypothermia—A serious loss of body heat resulting from being exposed to severe cold temperatures (p. 309)

Ice—A form of an amphetamine that is smoked (p. 204)
Immune—Resistant to infection (p. 236)
Immune system—A system of organs, tissues, and cells that fight infection (p. 235)
Immunization—Means of making a body immune from a disease (p. 236)
Income—The amount of money received or earned from a job (p. 379)
Infection—A sickness caused by a pathogen in the body (p. 233)
Infectious—Contagious (p. 306)
Infectious disease—A disease caused by a pathogen (p. 233)
Inflammatory response—The body's first response to a pathogen (p. 235)
Inhalant—A substance that is breathed (p. 206)
Inhalation poisoning—A poison that is inhaled (p. 318)
Inherited disease—A disease passed through genes (p. 232)
Inject—Use a needle to take medicine into the body (p. 190)

Inpatient—Someone receiving health care who stays in a facility overnight or longer (p. 350)
Internal conflict—A situation in which a person doesn't know what to do (p. 325)
Internet—The worldwide computer network that provides information to users (p. 291)
Intoxicated—Excited or stimulated by a drug (p. 199)
Involuntary muscle—A muscle that moves whether you think about it or not (p. 31)
Irregular—Not normal (p. 315)

Joint—A place where two bones come together (p. 28)

Kaposi's sarcoma—A rare type of cancer affecting the skin or internal organs (p. 244)
Keratin—A protein that makes nails hard (p. 26)

Landfill—A place where waste is buried between layers of earth (p. 396)
Large intestine—The tube that connects the small intestine and the rectum (p. 33)
Life-threatening emergency—Any situation in which a person might die if medical treatment isn't provided immediately (p. 292)
Look-alike drug—An illegal manufactured drug that imitates the effect of other drugs (p. 208)
Lot number—A number that identifies a group of packages (p. 167)
Lung cancer—A disease of the lungs caused primarily by smoking tobacco (p. 195)

Malignant tumor—A mass of cells that are harmful (p. 259)
Malnutrition—A condition in which the body does not get enough to eat (p. 162)
Marijuana—An illegal drug from the hemp plant that produces intoxication (p. 207)
Media—Sources of information and entertainment, such as newspapers and TV (p. 323)
Mediator—A person who helps two sides solve a problem reasonably (p. 332)

Medicaid—Health insurance for people with low incomes (p. 353)
Medicare—Health insurance for people age 65 or older or who receive Social Security disability (p. 353)
Medicine—A drug used to treat or prevent a disease or health problem (p. 186)
Medulla—The part of the brain that connects to the spinal cord (p. 40)
Melanin—A chemical that gives skin and hair its color (p. 26)
Menstruation—The monthly flow of an egg and extra tissue from the uterus (p. 48)
Metabolism—The process of cells using nutrients for energy and other needs (p. 140)
Metastasize—Spread cancer to distant tissues (p. 259)
Microscope—A tool used to see cells (p. 24)
Mineral—A nutrient from the earth that is needed to help the body use energy from other nutrients (p. 147)
Minor—A person under age 18 or 21 (p. 360)
Mitosis—The dividing process that makes new cells (p. 25)
Mucous membrane—The thin, moist tissue that lines body openings (p. 233)
Mucus—The sticky fluid produced by mucous membranes (p. 233)
Multiply—Increase in number (p. 233)
Muscular dystrophy—An inherited disease in which the muscles do not develop normally (p. 232)
Muscular system—The body system made up of your muscles (p. 28)

Narcotic—A drug that dulls the senses or relieves pain (p. 202)
Natural disaster—A destructive event that happens because of natural causes (p. 297)
Need—Something important or necessary to have (p. 93)
Nervous system—The body system that sends and receives messages throughout the body (p. 40)
Neutral—Not favoring either side (p. 333)
Nicotine—A chemical in tobacco to which people become addicted (p. 194)
Nucleus—The control center of the cell (p. 24)
Nurse practitioner—A registered nurse with special training for providing health care (p. 349)

Nutrient—The basic unit of food that the body can use (p. 140)
Nutrition Facts—The part of a food label that tells about the calories and nutrients in the food (p. 168)

O

Obesity—A condition in which one is more than 20 percent overweight (p. 162)
Ointment—A medicine for minor skin infections (p. 187)
Ophthalmologist—A doctor who specializes in diseases of the eye (p. 58)
Opiate—A drug made from opium poppy plants; another name for narcotic (p. 202)
Optometrist—A specialist in eye examinations and corrective lenses (p. 57)
Oral poisoning—A poison that is eaten or swallowed (p. 317)
Organ—A group of tissues that work together (p. 25)
Orthodontist—A dentist who treats crooked or crowded teeth (p. 57)
Osteoarthritis—A type of arthritis that causes a person's joints to get worse with age (p. 269)
Out-of-pocket—Straight from a person's income or savings (p. 352)
Outpatient—Someone receiving health care without staying overnight (p. 350)
Ovary—The female organ that stores eggs (p. 47)
Over-the-counter medicine—A medicine that can be bought without a doctor's written order (p. 186)
Ovulation—The monthly process of releasing an egg (p. 47)
Ozone layer—A region in the atmosphere that protects the earth from the sun's harmful rays (p. 392)

P

Particulate—A small pollutant, such as dust, ash, or dirt (p. 391)
Pathogen—A disease-causing germ (p. 232)
Pediatrics—Child health care (p. 350)
Peer—A person in the same age group (p. 108)
Peer pressure—The influence people of the same age have on one another (p. 108)
Penicillin—An antibiotic used to treat diseases such as gonorrhea (p. 248)
Penis—The male organ used to deliver sperm and to urinate (p. 49)

Peripheral nervous system—All the nerves in the body outside the brain and the spinal cord (p. 42)
Perspiration—Sweat (p. 26)
Petit mal seizure—A seizure that affects a person's mental functions (p. 271)
Pharmacist—A person trained and licensed to prepare and sell prescription drugs (p. 186)
Physical fitness—The body's ability to work, exercise, and play without tiring (p. 59)
Pituitary gland—The endocrine gland attached to the base of the brain (p. 46)
Plaque—A layer of bacteria on teeth (p. 56)
Plasma—The liquid part of blood (p. 38)
Pollution—The buildup of harmful wastes in the air, land, or water (p. 387)
Polyunsaturated fat—A fat that is found mostly in plant foods (p. 145)
Prejudice—An opinion based on a person's religion, race, gender, or culture (p. 328)
Premium—An amount of money paid to an insurance company at regular intervals (p. 352)
Prescription—A written order from a doctor for a medicine (p. 186)
Preservative—A chemical added to food to prevent spoiling (p. 171)
Prevention—Appropriate ongoing self-care (p. 348)
Primary care physician—A doctor who treats people for routine problems (p. 349)
Private resource—A group of people not associated with the government (p. 373)
Protein—A nutrient needed for growth and repair of body tissues (p. 142)
Psychedelic drug—Another term for hallucinogen (p. 205)
Psychoactive medicine—A medicine that changes the function of the brain (p. 188)
Puberty—The period when children develop into adults and reach sexual maturity (p. 47)
Public resource—A health service that is run by a local, state, or national government (p. 373)

Q

Quackery—A medical product or service that is unproved or worthless (p. 355)

R

Rabies—A disease transmitted to humans through animal bites (p. 310)
Radiation—A type of treatment that uses energy waves to destroy cancer cells (p. 261)

Radon—A colorless, odorless, poisonous gas that is formed underground (p. 391)
Recall—An order to return unsafe items (p. 375)
Rectum—The organ that stores solid waste before it leaves the body (p. 33)
Rehabilitation—Help to recover from surgery, illness, or injury (p. 350)
Relationship—A connection between people (p. 125)
Reproductive system—The body system responsible for making a baby (p. 47)
Resist—Act against (p. 109)
Resource—A source of supply and support (p. 370)
Respiratory system—The body system responsible for breathing (p. 36)
Rheumatoid arthritis—A type of arthritis caused by a defect in the immune system (p. 268)
Rheumatoid factor—The antibody associated with rheumatoid arthritis (p. 268)
Risk factor—A trait or habit that increases a person's chances of having or getting a disease (p. 258)

S

Saliva—The liquid in the mouth that begins digestion (p. 32)
Saturated fat—A fat that is found mostly in animal products (p. 145)
Screening service—An exam that provides prevention, such as eye and ear exams (p. 374)
Secondhand smoke—Tobacco smoke breathed by nonsmokers (p. 195)
Secrete—Form and give off (p. 45)
Seizure—A physical or mental reaction to disordered brain activity (p. 270)
Self-esteem—The way a person feels about himself or herself (p. 104)
Separation—A period when a married couple stops living together (p. 71)
Septic arthritis—A swelling of the joints caused by an infection (p. 269)
Serving size—A way to measure the amount of different foods that should be eaten each day (p. 152)
Sexual abuse—Any sexual contact that is forced on a person (p. 73)
Sexual intercourse—Inserting the penis into the vagina (p. 49)
Sexually transmitted disease—A disease spread by sexual contact (p. 216)
Shame—An emotion that results from disapproval or rejection (p. 91)

Shock—The physical reaction to injury in which the circulatory system fails to provide enough blood to the body (p. 311)
Shyness—Feelings of discomfort around others (p. 127)
Side effect—An unexpected and often harmful result of taking medicine (p. 191)
Skeletal system—The body system made up of your bones and joints (p. 28)
Small intestine—Where most of digestion takes place (p. 32)
Small talk—Talk about things that are interesting but not important (p. 123)
Smog—Air pollution formed by car exhaust and other pollutants (p. 390)
Smokeless tobacco—Tobacco that is chewed (p. 195)
Social emotions—Emotions that have to do with relationships with others (p. 90)
Specialist—A doctor who works only on certain types of medical problems (p. 346)
Sperm—The male sex cell (p. 48)
Spinal cord—The cable of nerve cells within the bones of the spine (p. 40)
Splint—A rigid object that keeps a broken limb in place (p. 307)
Sponsor—A recovering alcoholic who helps a new AA member (p. 218)
Sprain—The sudden tearing or stretching of tendons or ligaments (p. 307)
Statistics—Important information (p. 374)
Sterile—Unable to have children (p. 207)
Steroid—A chemical that occurs naturally in the body or is made in a laboratory (p. 207)
Stimulant—A drug that speeds up the central nervous system (p. 194)
Stress—The way the body reacts to a change or to something that can hurt you (p. 87)
Stress response—The body's reaction to stress (p. 105)
Stroke—A condition in which the blood supply to a person's brain is suddenly blocked (p. 257)
Subcutaneous layer—The deepest layer of the skin (p. 27)
Support group—A group of people with similar problems who help one another (p. 218)
Suppository—A cylinder containing medicine to be inserted into the rectum (p. 190)
Synthetic—A narcotic drug that is manufactured in laboratories (p. 202)
Syphilis—A sexually transmitted disease that has three stages (p. 249)

Tar—A substance in tobacco that can form a thick, brown, sticky substance in the lungs (p. 195)

Terminally ill—Dying from a disease, an injury, or an illness, sometimes over a long period (p. 351)

Testis—The male organ that makes sperm (p. 48)

Testosterone—The male hormone that produces male characteristics, such as facial hair and a deep voice (p. 207)

Therapeutic effect—A helpful result of taking medicine (p. 191)

Threat—A situation that puts a person's well-being in danger (p. 104)

Thunderstorm—A severe weather condition that produces thunder, lightning, and rain (p. 299)

Thyroid gland—The endocrine gland that affects a person's energy (p. 46)

Tissue—A group of cells that do the same job (p. 25)

Tolerance—A condition in which a person must take more and more of a drug to get the same effect (p. 202)

Tornado—A whirling, funnel-shaped storm that forms over land (p. 300)

Tornado warning—An alert issued when a tornado has been spotted in an area (p. 300)

Tornado watch—A situation in which a tornado may develop (p. 300)

Toxin—A dangerous chemical (p. 389)

Trachea—The tube that connects the throat to the lungs (p. 36)

Transfusion—The transfer of blood from one person to another (p. 245)

Transmit—Spread (p. 310)

Tremor—Severe shaking (p. 203)

Type I diabetes—Insulin-dependent diabetes (p. 265)

Type II diabetes—Non-insulin-dependent diabetes (p. 266)

Universal Precautions—The methods of self-protection that prevent contact with blood or other body fluids (p. 306)

Ureter—The tube that carries urine from the kidney to the bladder (p. 34)

Urethra—The tube through which urine passes out of the body (p. 34)

Urine—The liquid waste formed in the kidneys (p. 34)

Uterus—The female organ that holds a growing baby (p. 48)

Vaccination—An injection of dead or weakened viruses to make the body immune to the virus (p. 236)

Vaccine—A medicine that stimulates the immune system to fight off a disease (p. 188)

Vagina—The birth canal through which a baby is born (p. 48)

Vein—A blood vessel that carries blood back to the heart (p. 38)

Violence—Actions or words that hurt people or things they care about (p. 323)

Violent—Doing actions or words that hurt people (p. 72)

Vitamin—A nutrient needed to help the body use energy from other nutrients (p. 147)

Voluntary muscle—A muscle that moves when you think about it (p. 31)

Volunteer resource—A group of people who donate their time to provide services (p. 373)

Withdraw—Pull away (p. 102)

Withdrawal—A physical reaction to the absence of a drug (p. 196)

Withdrawal symptoms—The body's physical reaction to the absence of a drug (p. 214)

Word of mouth—Information about a product that you hear from a friend or family member (p. 355)

Index

A

Absorption
  of medicines, 190
  of nutrients, 140
Abuse
  alcohol, 199
  child, 73, 329–330
  drug, 202–3, 204, 205, 206, 207, 208, 245
  emotional, 73
  in families, 72–73
  sexual, 73
Acceptance, 72
Acid rain, 391
Acne, 55
Acquired disease, 232
Adapting, 89
Additives, 171
Adjustment, 72
Adolescents, 69
Adoption, 69
Adrenal gland, 46
Adrenaline, 46
Advertisements
  evaluating, 357–360
  of foods, 165–166, 178
  for harmful drugs, 360
  for health care products, 359
  on the Internet, 358
  methods of, 357–358
Advocacy groups, 380
Aerobic exercises, 59
Aggression, 101
Aging, effect of, on family, 69–70
Agricultural runoff in water pollution, 394
Agriculture, U.S. Department of, 172
AIDS (acquired immunodeficiency syndrome), 233, 242, 243. *See also* Sexually transmitted diseases
  detection of, 244
  myths about, 246
  risk of, from contaminated needles, 202
  spread of, 244
  symptoms of, 244
  transmission of, 245–246
Air, need for clean, 389
Airplanes, stopping violence in, 331
Air pollution
  causes of, 389
  effect of, on health, 389–91
  impact of, on environment, 391–392
Alateen, 218

Alcohol, 198, 284
  abuse of, 199
  combining with medicines, 199
  dangers in using, 199
  definition of, 198
  as depressant, 202
  driving under the influence of, 199–200
  effects of, on body, 198–99
  ethyl, 200
  harmful effects of, 199
  reasons for using, 198
  and safe use of, 199–200
  and violence, 328
Alcoholic beverages, 198
  safety of, 201
Alcoholics Anonymous (AA), 201, 218
  sponsors in, 218
Alcoholism, 200
  getting help, 201
Alcohol testing, software for, 200
Allergists, 349
Alternative birthing centers, 353
American Cancer Society, 373
American Heart Association, 309, 315
American Hospital Association, 362
American Red Cross, 308, 309, 315, 373
Amino acids, 144
Amphetamines, 204
Anabolic steroids, 207
Anger, 72, 88–89
  managing, 331
Animal bites, first aid for, 310
Anorexia, 121, 161, 162
Anti-Abuse Drug Act (1986), 208
Antibiotics, 187
Antibodies, 188, 236, 237
Antihistamines, 187
Anus, 33
Anxiety, dealing with, 106–107
Aorta, 38
Arteries, 38, 59
  diseases of, 256
Arteriosclerosis, 256
Arthritis
  definition of, 268
  osteoarthritis, 269
  rheumatoid, 268–269
  septic, 269
Asbestos, 391
Asthma, 263, 390
  treatment of, 264
  triggers for, 263
Atherosclerosis, 256
Automobiles, making safer, 286

B

Baby, effect of medicines on unborn, 192
Baby-sitters, safety for, 291
Banting, Frederick, 45
Barbiturates, 202–203
Barton, Clara, 308
Behavior
    changing, 89
    and emotions, 93–96
    violent, 323
Benign tumors, 259
Better Business Bureau, 363
Biofeedback, 107
Birth defects, preventing, 192
Black Death, 235
Blackheads, 55
Bladder, 34, 35
Bleeding, first aid for severe, 311
Blindness, 267
Bloating, 162
Blood, functions of, 37–38
Blood clots, 256
Blood pressure, 59, 61
    high, 255, 267
Blood tests
    in detecting AIDS, 244
    in detecting syphilis, 249
The Bluest Eye (Morrison), 94
Body chemistry and drug reactions, 191
Body image, 112
Body systems, 23. *See also* specific body systems
    definition of, 25
Bogus, 356
Bond, 119
Brady Handgun Violence Protection Act, 338
Brain
    functions of, 40–41
    parts of, 40
Breast cancer, 50, 261
Breasts, self-exam of, 50
Breath analyzers, 200
Bronchitis, 390
Bubonic plague, 235
Budget, 378
Bulimia, 113, 161, 162
Burns
    degree of, 308
    first aid for, 308
By-product, 389

C

Caffeine, 204, 284
Calcium, 149
Calories, 61, 141, 167
Cannabis sativa, 207

Cancer, 259
    breast, 50, 261
    lung, 195, 261
    prostate, 261
    self-exam for early signs of, 50
    skin, 261
    survival rates from, 260
    symptoms and warning signs of, 259
    testes, 50
    treatment for, 262
    types of, 260–261
Capillaries, 36, 38
Carbohydrates, 142–143
    complex, 143
    simple, 142
Carbon dioxide, 36, 38, 385
Carbon monoxide, 391
Cardiac arrest, 315
Cardiopulmonary resuscitation (CPR), 315–317, 318
Cardiovascular diseases, 188
    arteriosclerosis, 256
    atherosclerosis, 256
    definition of, 255
    heart attacks, 257
    and high blood pressure, 255
    risk factors for, 258
    strokes, 257
Cardiovascular fitness, 59
Cardiovascular medicines, 188
Careers
    dietary aide, 163
    drug abuse counselor, 220
    environmental protection worker, 397
    fitness instructor, 61
    food technologist, 141
    health service coordinator, 234
    home health aide, 378
    industrial safety specialist, 287
    laboratory assistant, 250
    medical records technician, 359
    medical specialists, 349
    mental health assistant, 95
    occupational therapy assistant, 257
    pharmacy clerk, 190
    physician's assistant, 334
    registered nurse, 314
    resident assistant, 114
    residential counselor, 124
Caries, 56
Cataracts, 267
Cell membrane, 24
Cells, 24
    metabolism in, 141
    parts of, 24–25
Center for Independent Living (CIL), 373
Centers for Disease Control (CDC), 374
Central nervous system, 42

Cerebellum, 40, 41
Cerebrum, 40
Charitable organizations, 376
Chemical replacements, 188
Chemicals, addition to foods of, 171–172
Chemotherapy, 262
Child abuse, 73, 329–330
Childishness, 101
Children, impact of separation and divorce on, 71
Chlamydia, 249
Chlorofluorocarbons, 392
Choking, 291
    first aid for, 312–314
Cholesterol, 145–146
Chronic disease, 268
Chronic infection, 247
Cigarettes. *See* Tobacco
Circulatory system, 140
    definition of, 36
    functions of, 37–38
    parts of, 38
Clinics, 371
Cocaine, 204
Cocaine Anonymous, 218
Codeine, 191
Commitment, 67
Common association, 369
Common identity, 369
Communicable diseases, 245, 370
Community
    changes affecting health of, 372
    contributions to, 380
    definition of, 369
    economics as concern of, 371
    factors in identifying, 369–370
    finding services in, 377
    health as concern of, 370–371
    health resources in, 370, 373–376, 377
    resource management in, 380
Community health advocacy skills, 377–380
Competition, 359
Complete protein, 144
Complex carbohydrates, 143
Compressions, 317
Compromise, 95
Computer programs for diet management, 152
Concentration camps, 329
Conflict
    avoiding, 331
    as cause of violence, 327
    dealing with serious, 334
    definition of, 325
    internal, 325
    resolving, 332
Consumer advocates, 361
Consumer Bill of Rights, 361
Consumer health, 345–364
Consumer Product Safety Commission (CPSC), 363, 375
Consumer Reports, 357

Consumers
    being wise, 354–356
    correcting problems of, 362–364
    definition of, 354
    influence of quackery on, 355–356
    influences on choices of, 354–355
    rights of, 361–362
    sources of help for, 363
Contact poisoning, first aid for, 318
Contamination, 172, 174
Contraction of muscles, 28
Convulsions, 203
Cool-down, 60
Coping, 102–103
Coughing, 41
Counseling and drug dependence, 217
Count, 355
Crack cocaine, 204
Crime
    impact on young people of, 329–330
    violent, 325
Crisis intervention, 73
Culture, effect on food choices, 165
Cycle of violence, 326, 327, 330
Cytoplasm, 24

D

Daily values, 169
Day care, 70
Decibels, 57
Decongestants, 187
Deductibles, 352
Defective products, 354
Denial, 72
Dental caries, 56
Dental offices, 351
Dentist, 57, 351
Depressants, 202–203, 284
    alcohol, 198–199
Depression, 71, 92
Dermatologist, 56, 349, 351
Dermis, 26–27
Designated drivers, 200
Designer drugs, 208
Detoxification, 217
Diabetes, 45, 188, 265
    gestational, 266
    health problems associated with, 267
    type I, 265–266
    type II, 266
Dietary guidelines, 151
    differences in, 152
    Food Guide Pyramid as, 151, 152
    general, 152
    for pregnancy, 153
    serving size in, 152
    special, 154

Diet management, computer programs for, 152
Diet-related health problems, 161–162
Digestion, 140
Digestive system, 32
Disappointment, dealing with, 91
Discipline, 68
Diseases
    acquired, 232
    body's protection from, 235–237
    cardiovascular, 255–258
    causes of, 232
    communicable, 245, 370
    definition of, 232
    infectious, 233–234, 306
    inherited, 232
Disinfectant, 198
Divorce, 71
Doctor, 351
    questions to ask about, 348
Driving under the influence of alcohol, 199–200
Drowning, 291
Drug abuse, 213
    of anabolic steroids, 207
    costs of, 216
    of depressants, 202–203
    of designer drugs, 208
    of hallucinogens, 205
    of inhalants, 206
    of look-alike drugs, 208
    of marijuana, 207
    of narcotics, 202
    and sexually transmitted diseases, 216
    and spread of AIDS, 245
    of stimulants, 204
Drug abuse counselor, 220
Drug dependence, 212–213
    definition of, 213
    effects of, 215–216
    problems linked to, 216, 371
    signs of, 214
    solutions to, 217–220
    and withdrawal, 214
Drugs, 186
    advertising harmful, 360
    effect of
        on emotions, 224
        on safety, 284
    reactions to, 187
    and violence, 328

Earthquake
    actions during and after, 298–299
    preparing for, 297

Eating disorders, 112, 161
    anorexia, 112, 161, 162
    bulimia, 113, 161, 162
    overeating, 114, 161, 162
Ecology, 388
Economics as community concern, 371
Ecstasy, 205
Egg, 48
Elderly, care for, 70
Elevation, 307
E-mail, 294
Emergency, contacts in, 292
Emergency equipment, 295–296
Emergency kit, 295
Emergency Medical Service (EMS), 305, 307, 308, 311, 373
Emotional abuse, 73
Emotional health, 61
Emotions
    and behavior, 93–96
    causes of, 87–89
    definition of, 86
    effect of drugs on, 224
    mental side of, 87
    physical side of, 86
    social, 90–92
Emphysema, 195, 390
Endocrine glands, 45
Endocrine system
    definition of, 44
    functions of, 44–45
    glands in, 46
Energy, daily needs for, 159
Enrichment, 171–172
Environment
    achieving balance in, 385
    definition of, 385
    effect of, on health, 387
    effect on food choices, 164
    factors affecting balance of, 386–387
    future solutions to problems in, 398
    impact of air pollution on, 391–392
    keeping healthy, 388
    reducing stress on, 396
    role of government in protecting, 397
    role of individuals in protecting, 399
Environmental health, 384–400
Epidemic, 235, 370
Epidermis, 26
Epilepsy
    definition of, 270
    diagnosis of, 272
    seizures in, 270–272
    treatment of, 272
Equipment, emergency, 295–296
Esophagus, 32
Ethyl alcohol, 200
Excretory system, 32–35

Exercise
    aerobic, 59
    benefits from, 60–61
    proper methods of, 60
Expenses, 379
Extended families, 67
Eyes
    caring for, 57–58
    first aid for objects in, 308

F

Families, 66
    changes in, 127
    dealing with problems in, 72
    definition of, 67
    effect of aging on, 69–70
    effect of stress on, 71
    effects of separation and divorce, 71
    extended, 67
    foster, 69
    getting help with problems, 74
    growth of, 67–68
    impact of drug dependence on, 215
    violence and abuse in, 72–73, 324
Family counselors, 73
Family life cycle, 67
    stages in, 67–70
Famine, 395
Fats, 145
    polyunsaturated, 145
    saturated, 145
Fear, 88, 89
Febrile seizure, 272
Federal Trade Commission Office of Public Affairs, 363
Feelings, effect on food choices of, 163
Fiber, 143
Fight or flight response, 105
Fights, avoiding, 332–333
Fire, reducing risks of, 288–290
Firearms, 328. *See also* Gun control
    safety with, 287
Fire escape routes, planning, 290
First aid
    for animal bites, 310
    for burns, 308
    for choking, 312–314
    definition of, 305
    for fractures, 307
    for frostbite, 310
    guidelines for, 305
    for heart attacks, 315–317
    for heat exhaustion, 309
    for heatstroke, 309
    for hypothermia, 309
    for nosebleeds, 309
    for objects in the eye, 308
    for poisoning, 317–318
    for severe bleeding, 311
    for shock, 311–312
    for sprains, 307
    training in, 309
First aid kit, 296
Fitness
    cardiovascular, 59
    importance of rest and sleep to, 61–62
    physical, 59
    planning program in, 78
Fitness instructor, 61
Flashbacks, 205
Floods, 300
Flood warning, 300
Flood watch, 300
Food and Drug Administration (FDA), 171, 189, 363, 374
Food Guide Pyramid, 151–152
Food labels
    comparing, 170
    information on, 168–169
    reading, 167–170
    reasons for reading, 170
Food record, keeping, 152
Foods, 139. *See also* Nutrients
    additives in, 171
    advertising of, 165–166, 178
    contamination of, 172, 174
    digestion of, 140
    influences on choices of, 163–166
    junk, 153
    keeping nutrients in, 173
    making healthy choices in, 158–174
    preservatives in, 171
    proper handling of, 172
Food safety, 171–174
    agencies controlling, 171, 172
    and avoiding infectious disease, 234
    rules for, 174
    temperature in, 173
Food Safety and Inspection Service, 363
Food technologist, 141
Fortification, 171–172
Fossil fuels, 390
Foster families, 69
Fractures, 307
    first aid for, 307
Freshwater, conservation of, 395
Friendships, 119–121
    end of, 128
    forming, 123–124
    importance of values in, 122
    problems in, 127–128
    types of, 122
Frostbite, first aid for, 310
Frustration, managing, 101–103

G

Generic products, 356
Genes, as cause of disease, 232
Genital herpes, 247
Genocide, 329
Gestational diabetes, 266
Ginger, 35
Glands, 26, 44
    adrenal, 46
    endocrine, 45
    oil, 27, 55
    pituitary, 46, 48
    sweat, 27
    thyroid, 46
Glaucoma, 267
Glucose, 265
Gonorrhea, 248
Good Samaritan Laws, 306
Gossip, 128
Government, role of, in protecting environment, 397
Grand mal seizure, 270, 271
Grants, 375
Greenhouse effect, 392
Grief, 91–92
Groundwater, 393
Growth hormones, 46
Guilt, 90
Gum disease, 56
Gun control. *See also* Firearms
    in preventing violence, 338

H

Hair, taking care of, 56
Hallucinations, 205
Hallucinogens, 205
Hardening of the arteries, 256
Hassles, 104
Hazards, 288
Health
    as community concern, 370–371
    effect of environment on, 387
    effect of land pollution on, 395
    impact of drug dependence on, 215
    importance of water to, 393
Health care
    definition of, 346
    paying for, 352–353
    professionals assisting with, 346
    sources for, 351
Health care facilities, 350–351
Health care products, 347
    advertising, 359
Health care professionals
    finding, 350
    identifying, 349
Health care programs, funding for, 376
Health goals, identifying national, 371–372
Health inspectors, 172
Health insurance, 352
    government-supported, 353
    private, 352
Health maintenance organization (HMO), 351
Health resources, finding, in community, 377
Hearing
    caring for, 58
    protecting, 57
Heart, 59
Heart attacks, 59, 146, 257
    first aid for, 315–317
Heart disease, 59, 267
Heart rate, 59
Heat exhaustion, first aid for, 309
Heatstroke, first aid for, 309
Heimlich maneuver, 312–314
Heroin, 202
Herpes, genital, 247
High blood pressure, 59, 255, 267
HIV (human immunodeficiency virus), 243.
    See also AIDS (acquired immunodeficiency syndrome)
HIV negative, 244
HIV positive, 244
Hormones, 45, 55
Hospices, 245, 351, 353
Hospital, 351
Household chemicals in water pollution, 394
Hughes, Langston, 214
Human activity, effect of, on environment, 386
Hurricane Mitch, 386
Hurricanes, 299
Hurricane warning, 299
Hurricane watch, 299
Hydrocarbons, 389
Hygiene, 55
    dental care in, 56–57
    hair care in, 56
    nail care in, 56
    practicing good, 121
    skin care in, 55–56
Hypertension, 255
Hypothermia, first aid for, 309

I

Ice, 204
ICE (immobilize, cold, elevate), 307
Ignition interlock, 200
Immune, 236
Immune system, 235, 237, 243
    in fighting disease, 236
Immunization, 236–237
Income, 379
Indoor pollution, impact on health of, 391
Industrial wastes in water pollution, 393

Infection, 233
Infectious diseases, 233, 306
    avoiding, 234
Inflammatory response, 235
Inhalants, 206
Inhalation poisoning, first aid for, 318
Inherited diseases, 232
Injections, 190
Injuries
    first aid for, 305–318
    preventing, 263
Inpatient care, 350
Insulin, 45, 188, 265
Insulin-dependent diabetes, 265
Internal conflict, 325
International health resources, 375
Internet, 291
    advertising on, 358
    protecting yourself on, 293
Intoxication, 199
Involuntary muscles, 31, 41
Iron, 149
Irregular heartbeat, 315

Joints, 28
Joule, 167
"Junior Addict" (Hughes), 214
Junk foods, 153

Kaposi's sarcoma, 244
Keratin, 26
Kidnapping, 330
Kidneys, 35
    failure of, 59

Laerdal, Asmund S., 315
Landfill, 396
Land pollution, effect on health of, 395
Large intestine, 33
Laser surgery, 56
Lead poisoning, 391
Librium, 224
Life-threatening situations, 292, 311
Light, sources of, in emergency, 295
Lightning, 299
Listening, skills of, 132
Local health resources, 373–374
Long-term care facility, 351
Look-alike drugs, 208
Loss, dealing with, 91

Lot number, 167
Love, 90
LSD, 205
Lung cancer, 195, 261
Lungs, 38

Malignant tumors, 259
Malnutrition, 162
Managed care, 353
Marijuana, 207, 284
Marriage, 67
    characteristics of healthy, 68–69
MDMA, 205
Measles vaccination, 236
Mediator, 332
Media violence, 323
Medicaid, 351, 353
Medical care
    and appropriateness of self-care, 348
    seeking professional, 348
    sources of, 347
Medical offices, 351
Medical specialists, 346, 349
Medicare, 351, 353
Medicines, 183, 185
    administration of, 190
    benefits of, 185–189
    combining alcohol with, 199
    different reactions to, 191
    effect of, on unborn baby, 192
    effects of, on body, 191
    over-the-counter, 185, 346
    patent, 189
    prescription, 185
    reactions to, 187
    rules for taking, 193
    safety of, 189
Medulla, 40, 41
Melanin, 26
Menstruation, 48, 154
Mental health, maintaining, 100–114
Mental health assistant, 95
Mental health counselors, 95, 217
Mental illness, treatment of, 113
Metabolism, 141
Metastasis, 259
Metric units, 167
Microscope, 24
Minerals, 147, 148–150
Minors, 360
Mitosis, 25
Mothers Against Drunk Driving (MADD), 200
Motor vehicle safety rules, 285
Mucous membranes, 233
Mucus, 233

Muscles
    contraction of, 28
    involuntary, 31, 41
    voluntary, 31, 41
Muscular dystrophy, 232
Muscular system, 28

Nails, taking care of, 56
Narcotics, 202
Narcotics Anonymous, 218
National Council on Alcoholism and Drug
    Dependence, 218
National health resources, 374–375
National Institutes of Health (NIH), 375
Natural disasters
    results of, 386
    safety during, 297–300
Needs, 93
    in relationships, 126
Neighborhood health clinics, 353
Nerve disease and diabetes, 267
Nerves, 42–43
Nervous system
    definition of, 40
    function of, 40
    parts of, 42–43
    role of brain in, 40–41
Neutral, 333
Nicotine, 194, 196
911 software, 306
911 telephone calls, 292, 305
Noise pollution, 58
Nonsmokers, rights of, 196–197
Nosebleeds, first aid for, 309
Nucleus, 24–25
Nurse practitioner, 349, 351
Nutrients, 140. *See also* Foods
    absorption of, 140
    carbohydrates as, 142–143
    fats as, 145–145
    keeping, in food, 173
    minerals as, 148–150
    proteins as, 143–144
    vitamins as, 147–148
    water as, 150
Nutrition Facts, 168

Obesity, 162
Obstetricians, 349
Occupational Safety and Health Administration
    (OSHA), 397
Oil glands, 27, 55

Oil leaks and spills in water pollution, 394
Ointments, 187
Oncologists, 349
Opiates, 202
Ophthalmologists, 58, 349
Optometrist, 57–58
Oral poisoning, first aid for, 317
Orderlies, 43
Organ, 25
Orphanages, 69
Orthodontist, 57
Orthopedic surgeons, 349
Osteoarthritis, 269
Outdoor pollution, impact on health of, 389–390
Out-of-pocket expenses, 352
Outpatient care, 350, 353
Outpatient clinic, 351
Outpatient treatment centers, 219
Ovaries, 47
Overeating, 114
Over-the-counter medicines, 186, 346
Ovulation, 47
Oxygen, transport of, through body, 36, 38
Ozone, 390, 392

Pain medicines, 186, 188
Particulates, 391
Patent medicines, 189
Pathogens, 232
    and avoiding infectious disease, 234
    body's methods of stopping, 233
    role of body in fighting, 235
Patient's Bill of Rights, 362
PCP, 205
Pediatricians, 349
Pediatrics, 350
Peer pressure
    definition of, 108
    resisting, 110–111
    responding to, 108–109
Peers, 108
Penicillin, 248
Penis, 49
People, effect on food choices of, 165
Peripheral nervous system, 42–43
Personal health habits, 276
Perspiration, 27, 55
Petit mal seizure, 271
Pharmacist, 186
Pharmacy clerk, 190
Phencyclidine (PCP), 205
Physical fitness, 59
Physical health, 61
Physical therapist, 351
Pimples, 55

Discover Health Index **429**

Pituitary gland, 46, 48
Plague, 56
Plasma, 38
Poison control center, 317
Poisoning, 291
    first aid for, 317–318
    lead, 391
Pollution, 387
    air, 389–392
    land, 395
    noise, 58
    water, 393–394
Polyunsaturated fats, 145
Population changes, 379
Population growth, 395
    and need for clean water supply, 404
Poverty, as community concern, 371
Preferred provider organizations (PPO), 351
Pregnancy
    dietary guidelines for, 153
    effect of medicines during, 192
    effects of smoking on, 195
    and gestational diabetes, 266
Prejudice, 328
Premium, 352
Prescription, 186
Prescription medicines, 186
Preservatives, 171
Prevention, 348
Primary care physician, 349
Private health insurance, 352
Private resources, 373
Products, judging, 356
Prostate cancer, 261
Proteins, 142, 143
    complete, 144
    lack of, 162
Psychedelic drugs, 205
Psychiatrist, 351
Psychoactive medicines, 188–189
Puberty, 47, 48
Public health, 368–380
Public Health Department, 172
Public Health Service, 235
Public resources, 373

Q

Quackery, 355–356, 361

R

Rabies, 310
Radiation
    and skin cancer, 261
    in treating cancer, 262

Radio, 295
Radon, 391
Random violence, 324
Recall, 375
Recipes, improving, 161
Rectum, 33
Recycling, 400
Red blood cells, 38, 149
Registered dietitian, 351
Rehabilitation, 350
Relationships, 118–128. *See also* Friendships
    definition of, 125
    end of, 128
    healthy, 125–126
    needs in, 126
    rights in, 125
    values in, 126
Reproductive system
    and creation of baby, 49
    definition of, 47
    female
        and birth of baby, 48
        parts of, 47–48
        and release of egg, 48
    male
        parts of, 48
        and release of sperm, 49
Resident assistant, 114
Residential counselor, 124
Residential treatment centers, 219
Resistance, 109
Resources, 370
    community management of, 380
    health, 377
    international health, 375
    local health, 373–374
    management of, 378
    national health, 374–375
    private, 373
    public, 373
    volunteer, 373
Respiratory system
    definition of, 36
    functions of, 36
Restaurants, effect on food choices of, 164
Resuci-Annie, 315
Rheumatoid arthritis, 268–269
Rheumatoid factor, 268
Rights in relationships, 125
Risk factor, 258
    cardiovascular, 258
Running shoes, 31

Safety
    automobile, 285, 286
    for baby-sitters, 291

    effect of drugs on, 284
    fire, 288–290
    with firearms, 287
    impact of drug dependence on, 215
    of medicines, 189
    during natural disasters, 297–300
    promoting, 284–287
    sports, 286
    for teens, 291–294
Saliva, 32, 140
Saturated fats, 145
Schools as local health resources, 374
Screening services, 374
Seasonal affective disorder (SAD), 105
Secondhand smoke, 195, 196–197
    avoiding, 197
Secretion of hormones, 45
Seizures, 272
    epileptic, 270
    grand mal, 270, 271
    petit mal, 271
Self-care, appropriateness of, 348
Self-esteem, 104
Self-exam
    of breasts, 50
    of testes, 50
Self-talk, 120
Separation, 71
Septic arthritis, 269
Services, judging, 356
Serving size, 152
Sewage in water pollution, 393
Sexual abuse, 73
Sexual activity, saying "no" to, 248
Sexual contact and spread of AIDS, 245
Sexual intercourse, 49
Sexually transmitted diseases, 242. *See also* AIDS
    (acquired immunodeficiency syndrome)
    AIDS as, 247
    chlamydia as, 249
    definition of, 247
    and drug abuse, 216
    genital herpes as, 247
    gonorrhea as, 248
    prevention of, 249–250
    syphilis as, 249
Shame, 91
Shock, first aid for severe, 311–312
Shoes, running, 31
Shyness, 127
Side effects, 191
Simple carbohydrates, 142
Skeletal system, 28
Skin
    function of, 26
    problems in, 55–56
    structure of, 26–27
    taking care of, 55
Skin cancer, 261

Sleep, importance of, in fitness, 61–62
Small intestine, 32, 140
Small talk, 123
Smog, 390
Smoke detectors, 288, 289
Smokeless tobacco, 196–197
Snacks, choosing healthy, 159–160
Snakebites, first aid for, 310
Sneezing, 41
Social emotions, 90–92
Social health, 61
Social studies, 245
Society, impact of drug dependence on, 216
Sodium, 150
Software for alcohol testing, 200
Solid waste, 396
Specialist, 346
Sperm, 48
    release of, 49
Spinal cord, 40, 41
Splint, 307
Sports, women in, 62
Sports-related injuries, 286
Sprains, 307
    first aid for, 307
State health resources, 374
Statistics, 374
Sterility, 207
Steroids, 207
Stimulants, 204, 284
    definition of, 194
    nicotine as, 194
Stomach, 140
Stress, 87, 94
    causes of, 104
    impact on family, 71
    managing, 104–106
    physical signs of, 105
Stress management, 61
Stress response, 105
Strokes, 59, 257, 267
Students Against Drunk Driving (SADD), 200
Subcutaneous layer, 27
Sugar, 142
Suicide and drug dependence, 215
Sunscreens, 55
Sun protection factor (SPF), 55
Support, 218
Support groups, 218, 373
Support system, creating own, 121
Suppository, 190
Surgery
    laser, 56
    in treating cancer, 262
Swallowing, 41
Sweat glands, 27
Symptom relievers, 187–188
Synthetic narcotics, 202
Syphilis, 249

Discover Health   Index **431**

Tars, 195
TDD (telecommunications device for the deaf), 306
Teens, safety for, 291–293
Teeth, taking care of, 56–57
Terminal care, 351
Testes, 48
    cancer of, 50
    self-exam of, 50
Testosterone, 207
Threats, 104
Thunderstorm, 299
Thyroid gland, 46
Tissue, 25
Tobacco, 194
    effects of, 194, 195, 219
    and lung cancer, 261
    reasons for using, 194
    and rights of nonsmokers, 196–197
    and secondhand smoke, 195, 196–198
    stopping use of, 196
Tolerance, 202, 213
Tonsillitis, 236
Tonsils, 236
Tornado, 300
Tornado warning, 300
Tornado watch, 300
Toxins, 389
Trachea, 36
Tranquilizers, 202–203
Transfusions and spread of AIDS, 245
Transmission, 310
Tremors, 203
Tumors
    benign, 259
    malignant, 259
Type I diabetes, 265
Type II diabetes, 266

United Way, 376
Universal Precautions, 306
Ureters, 34, 35
Urethra, 34, 35, 49
Urine, 34, 35
Uterus, 48

Vaccinations, 236–237
    tracking, 238
Vaccines, 188
Vagina, 48, 49
Valium, 224

Values
    importance of, 122
    in relationships, 126
Veins, 38, 59
Violence
    causes of, 327–329
    versus conflict, 324
    conflict as cause of, 327
    costs of, 326
    and crime, 329–330
    cycle of, 326, 327, 330
    definition of, 323
    and drug dependence, 215
    effects of, 325
    in families, 72–73, 324
    family, 324
    and genocide, 329
    gun control in preventing, 338
    media, 323
    preventing, 331–334
    random, 324
Violent behavior, 323
Vitamins, 147–48
Voluntary muscles, 31, 41
Volunteer resources, 373

Warm-up, 60
Wastes
    removal of liquid, from body, 34–35
    removal of solid, from body, 33
Water, 147, 150
    importance of, to health, 393
    importance of clean, 404
Water pollution, causes of, 393–394
Weight
    calculating ideal, 154
    maintaining healthy, 160
Westermarck, Alexander, 127
Wetlands, disappearing, 399
Wheezing, 263
White blood cells, 24, 38
White Cross Electric Vibrator Chair, 364
Withdrawal from drugs, 196, 214
Withdrawing, frustration and, 102
Women in sports 62
Word of mouth, 355
World Health Organization (WHO), 375

Young people, impact of crime on, 329–330